FRCOphth Part 2: MCQs

Julian Tagal MBChB FRCOphth
Senior Registrar in Ophthalmology
Sarawak General Hospital
Malaysia

Hee Ninh Ling MD MRCP
Senior Registrar in Dermatology
Sarawak General Hospital
Malaysia

ISBN 978-1-312-06628-1
90000

9 781312 066281

Table of Contents

Paper 1 7

Paper 1: Answers and Discussion 21

Paper 2 37

Paper 2: Answers and Discussion 51

Paper 3 67

Paper 3: Answers and Discussion 81

Paper 4 97

Paper 4: Answers and Discussion 111

Paper 5 127

Paper 5: Answers and Discussion 141

Preface

We wrote this book in order to primarily aid Trainee Ophthalmologists preparing for the Final FRCOphth written exam. The FRCOphth exam has undergone various changes in format over the past few years, and the Part 2 currently consists of two Multiple Choice Question (MCQ) papers, an OSCE and a Viva.

While there are several well-written preparatory books on the market, they consist mainly of Extended Matching Questions (EMQ) or True or False Questions.

We wanted to write this book in order to best reflect the current Final FRCOphth format in the hope that candidates would be able to best prepare for the examination, especially first time candidates with no prior experience.

This book consists of 5 MCQ papers, consisting of 90 questions each to reflect the current exam format. A chapter consisting of answers, discussion and references then follows each paper.
We have tried our best as possible to ensure that the proportion of questions allocated to each Ophthalmic Subspecialty, Therapeutics, Investigations, Royal College/NICE Guidelines and Miscellaneous topics closely reflects the question proportions as stated in the official Examination Reports available on the College website at rcophth.ac.uk.

We have also tried our best to ensure that the style of questions closely reflects that of the actual exam

Whilst we wrote this book to cater to the needs of candidates sitting for the FRCOphth examination, we hope that it will be equally useful for candidates who are preparing for the FRCS and various Masters examinations.

We enjoyed writing the book and we do hope it will serve you well as you prepare for your exams.

Julian Tagal
Hee Ninh Ling

Paper 1

1. Which of the following hereditary vitreo-retinal degenerations is autosomal recessively inherited?

a. Stickler's Syndrome
b. Familial Exudative Vitreo-retinopathy
c. Knobloch Syndrome
d. Wagner disease

2. Which of the following antibiotics is LEAST likely to be active against MRSA?

a. Vancomycin
b. Teicoplanin
c. Ciprofloxacin
d. Gentamicin

3. The following statements regarding astigmatism are TRUE except?

a. Oblique astigmatism is characterized by a steepest meridian that coincides with neither the vertical or horizontal meridians
b. Against the rule astigmatism is characterized by a steepest meridian that lies at 180 degrees
c. Is of the regular type in keratoconus
d. Children often have with-the-rule astigmatism

4. The following are TRUE about the clotting cascade except?

a. May be activated via an intrinsic or extrinsic pathway
b. The extrinsic pathway requires endothelial damage for activation
c. The intrinsic pathway appears to play an important role in the inflammatory process
d. Requires the presence of essential co-factors such as Vitamin C, Vitamin K and Calcium

5. Which of the following phakomatoses is the MOST common?

a. Tuberous Sclerosis
b. Neurofibromatosis Type 2
c. Von Hippel Lindau Syndrome
d. Sturge Weber Syndrome

6. The following viruses that are significant to the field of Ophthalmology are DNA viruses EXCEPT?

a. Herpes Simplex
b. Cytomegalovirus
c. Rubella
d. Molluscum contagiosum

7. A saddle nose may be seen in the following conditions EXCEPT?

a. Relapsing polychondritis
b. Herpes Zoster
c. Wegener's Granulomatosis
d. Syphilis

8. Which of the following phakomatoses is autosomal recessive?

a. Neurofibromatosis Type 1
b. Tuberous Sclerosis
c. Ataxia Telangiectasia
d. Von Hippel Lindau Syndrome

9. Regarding posterior polar cataracts, which of the following statements is MOST likely to be false?

a. Are associated with remnants of the vascular hyaloid system
b. Typically inherited as an autosomal recessive trait
c. Associated with an increased risk of posterior capsule rupture
d. The gene is mapped to the long arm of Chromosome 16

10. Which of the following is NOT a risk factor for conversion of ocular hypertension to primary open angle glaucoma (POAG)?

a. Higher intraocular pressure (IOP)
b. Thinner central corneal thickness
c. Greater Mean Deviation (MD)
d. Greater cup disc ratio

11. The following are TRUE about the natural history of central retinal vein occlusion (CRVO) except?

a. The prognosis in terms of visual acuity largely depends on the presenting visual acuity
b. 20% of patients presenting with a visual acuity of 6/60 or better improve spontaneously to 6/15 or better
c. 80% of patients presenting with a visual acuity of 6/60 or less do not improve or may even worsen
d. More than half of patients with an ischaemic CRVO present with neovascular glaucoma within 4 months

12. Regarding Kollner's rule, which statement is MOST likely to be false?

a. Blue – yellow colour vison defects arise as a result of macular disorders
b. Red – green colour defects arise as a result of optic nerve disorders
c. Glaucoma is an exception to Kollner's rule
d. Colour defects in cone dystrophies obey Kollner's rule

13. Which of the following is the MOST likely diagnosis for the following scenario?

A 70-year-old lady who is an ardent cyclist presents with an elevated, hyperkeratotic lesion on her left lower eyelid. There is central ulceration. Histopathology reveals keratin pearls and intercellular bridging.

a. Actinic Keratosis
b. Squamous Cell Carcinoma
c. Seborrhoeic Keratosis
d. Keratoacanthoma

14. Regarding orbital cellulitis in children, which of the following statements is MOST likely to be true?

a. The commonest causes are anaerobic bacteria of the upper respiratory tract
b. In temperate climates, the incidence of orbital cellulitis tends to be constant throughout the year
c. Affects both males and females equally
d. In very young children is almost always associated with ethmoid sinusitis

15. For the following scenario, which is MOST likely to be the correct diagnosis?

A 9-month-old baby girl presents with a large angle esodeviation. According to her mother, it has been there since birth. On examination, the esodeviation measures 45 prism diopters at distance and at near. It is alternating. Motility is full, with the angle of deviation similar in all directions of gaze. She appears to use her right eye to fixate in left lateral gaze and vice versa.

a. Accommodative Esotropia
b. Congenital Esotropia
c. Cyclical Esotropia
d. Decompensating Esophoria

16. The following statements regarding hyphaema are TRUE except?

a. Over half of patients with traumatic hyphaema have associated posterior segment injury
b. Elevated intraocular pressure may occur in up to a quarter of patients
c. Corneal bloodstaining does not occur in the presence of a normal intraocular pressure (IOP)
d. A black ball hyphaema is a risk factor for corneal bloodstaining

17. Which of the following conditions does NOT present with optically empty vitreous?

a. Jansen Syndrome
b. Wagner Syndrome
c. Marfan Syndrome
d. Goldman Favre

18. Which is likely to be the BEST culture media for the following scenario?

A 2-day-old baby girl presents with copious purulent discharge from both eyes. There is corneal infiltration. Her mother gives a history of multiple sexual partners.

a. Thayer Martin Agar
b. Chocolate Agar
c. Nutrient Agar with Klebsiella overlay
d. MacConkey

19. Which is likely to be the BEST investigation for the following scenario?

A 45-year-old gentleman presents with watery eyes. There is no ocular surface disease. He is not on any medication. There is no laxity of the periorbital tissues. Probing and syringing shows bilateral hard stops with presence of saline in his nasal passages. Primary Jones testing is negative, and secondary Jones testing is positive.

a. Computerised tomography (CT) scan of the midface
b. Dacryocystorhinogram
c. Lacrimal Scintigraphy
d. Repeat probing and syringing

20. A 45-year-old gentleman presents with sudden onset binocular diplopia. On examination his right eye has complete ptosis. The pupil is dilated and non reactive. He is unable to elevate, adduct or depress the eye. Abduction is intact. He has no diabetes or hypertension and is a non-smoker.

What would be the BEST way to proceed in investigation of this patient?

a. Contrasted Computed Tomographic (CT) scan of the brain
b. Cerebral Digital Subtraction Angiography (DSA)
c. CT Angiogram of the intracranial vessels
d. 1.5 Tesla Magnetic Resonance Angiography (MRA)

21. Regarding tonometers, which of the following statements is LEAST likely to be correct?

a. Goldman tonometry is based on the assumption that there is little variability between human eyes
b. A false diagnosis of normotension glaucoma may be made in patients with very thin corneas
c. Intraocular pressure may be falsely high in patients with recurrent herpes keratitis
d. Patients with decompensated Fuch's endothelial dystrophy may present with an artificially high intraocular pressure

22. The following statements regarding the role of contrast sensitivity testing are TRUE except?

a. Contrast sensitivity testing may be useful in detecting disease in patients with normal or near-normal acuity
b. Loss of contrast sensitivity typically occurs in patterns specific for different ocular pathologies
c. Has been measured as a secondary outcome in the Optic Neuritis Treatment Trial (ONTT)
d. The Vistech Chart is an example of contrast sensitivity tests utilizing sine wave gratings

23. A 49-year-old gentleman presents with proptosis, ophthalmoplegia and eye redness. He has shortness of breath and haematuria. What is the investigation of choice?

a. C –ANCA
b. P – ANCA
c. Antinuclear Antibodies
d. Rheumatoid Factor

24. According to the findings of the following investigations, what is the MOST likely diagnosis?

An elevated orange lesion with high internal reflectivity on ultrasound.
Fluorescein angiography reveals early hypofluorescence, absence of intra-lesional vessels, and the presence of late, pin- point hyperfluorescence.

a. Choroidal Melanoma
b. Choroidal Metastasis
c. Choroidal Haemangioma
d. Choroidal Melanocytoma

25. The following genetic markers are associated with a worse prognosis for uveal melanoma EXCEPT?

a. Chromosome 8q Gain
b. Monosomy 3
c. Chromosome 6p Gain
d. Mutations involving the gene for BRCA associated Protein 1 (BAP1)

26. The following statements regarding ocular penetration of topically administered drugs are FALSE except?

a. The corneal epithelium presents a major barrier to lipophilic molecules
b. The corneal stroma presents a major barrier to hydrophilic molecules
c. The sclera is similar to corneal stroma in terms of resistance to molecular movement
d. Normal basal tear secretion does not have any significant effect on drug contact duration

27. Which of the following statements regarding Punctate Inner Choroidopathy (PIC) is MOST likely to be false?

a. Is associated with myopia
b. Affects women more than men
c. There is typically no permanent structural damage to the retina
d. 30% of patients develop secondary choroidal neovascularization (CNV)

28. A 35-year-old lady with history of Steven Johnson Syndrome and severe dry eye presents to your corneal clinic. She has had bilateral corneal grafts that have failed twice in both eyes. Her eyes are extremely dry. Her visual acuity is counting fingers in both eyes. B scan of her eyes reveal no posterior segment abnormality.

What would be the treatment option of choice in this lady?

a. Penetrating keratoplasty (PK)
b. Osteo-Odonto-Keratoprosthesis (OOKP)
c. Phototherapeutic Keratectomy
d. EDTA chelation and scraping

29. Selective Laser Trabeculoplasty (SLT) is a surgical option for treatment of the following glaucomatous disorders EXCEPT?

a. Iridocorneal Endothelial Syndrome (ICE)
b. Primary Open Angle Glaucoma (POAG)
c. Pigment Dispersion Syndrome (PDS)
d. Pseudoexfoliation Syndrome (PEX)

30. Regarding Toxic Anterior Segment Syndrome (TASS), which of the following statements is MOST likely to be false?

a. May be due to preserved adrenaline added to balanced salt solution irrigation fluid
b. Typically presents with intense anterior segment inflammation within 12-24 hours post surgery
c. Is characterized by severe pain
d. Responds rapidly to intensive topical steroids

31. Which of the following statements regarding the diagnosis and management of Stevens Johnson Syndrome is MOST likely to be false?

a. The diagnosis is clinical
b. Conjunctival biopsies from inflamed conjunctiva may reveal sub-epithelial plasma cells and lymphocyte infiltration
c. Systemic corticosteroid therapy markedly improves survival rates
d. Anterior lamellar keratoplasties for corneal scarring should only be considered after disease quiescence of at least 3-6 months

32. What is the MOST likely diagnosis, according to the following scenario?

A 50-year-old lady who has been recently diagnosed with glaucoma presents with bilateral red eyes. On examination, she has chemosis and follicles in her inferior fornix. The skin around her eyes is excoriated, inflamed and sore. She denies any recent sexual activity.

a. Drop Hypersensitivity
b. Inclusion Conjunctivitis
c. Adenoviral Conjunctivitis
d. Allergic Conjunctivitis

33. Regarding steroid induced glaucoma, which of the following statements is MOST likely to be false?

a. There is no gender bias
b. It is a form of open angle glaucoma
c. Intravenous steroids carry a higher risk of inducing glaucoma as compared to inhaled steroids
d. Steroid response is seen in a larger proportion of eyes with primary angle closure (PAC) as compared to normal eyes

34. Which of the following macular disorders is the LIKELIEST diagnosis for the scenario below?

A 20-year-old lady presents with sudden onset blurring of vision in her left eye. She has a history of trauma to her left eye when she was still in school. Her visual acuity is OD: 6/36 OS: 6/60 unaided. She wears glasses that minify images. The fundi are tessellated. Her optic discs are tilted. The macula in the right eye is normal. A crescentic, circumferential scar is seen to involve the left macula, with a large sub retinal haemorrhage involving the parafoveal region.

a. Myopic Macular Degeneration with Choroidal Neovascularisation
b. Choroidal Rupture with secondary Choroidal Neovascularisation
c. Age Related Macular Degeneration
d. Retinal Angiomatous Proliferation

35. Of the following, which is the LEAST likely to be a life threatening complication associated with retinitis pigmentosa?

a. Cardiac conduction defects
b. Renal impairment
c. Liver cirrhosis
d. Raised intracranial pressure

36. Which of the following statements regarding Irvine Gass Syndrome (IGS) is MOST likely to be false?

a. Topical latanoprost does not have to be discontinued prior to cataract surgery in patients with glaucoma
b. Posterior capsular rupture is associated with an increased risk of Irvine Gass Syndrome
c. Fluorescein angiography typically reveals petalloid leakage involving the macula
d. Less than 50% of patients achieve a visual acuity of 6/9 or better

37. According to Royal College guidelines for Age Related Macular Degeneration, the following are contraindications for treatment with intravitreal ranibizumab EXCEPT?

a. Lesion size greater than 12 disc diameter
b. Visual acuity of worse than 6/60
c. Foveal scarring
d. Documented hypersensitivity to ranibizumab

38. Regarding disclosure for Commercial Sponsorship, in which of the following scenarios is financial disclosure NOT necessary?

a. Receipt of a USB stick, a DVD player and car cushions from a pharmaceutical company amounting to £105 for the past calendar year
b. A pen worth £15
c. Sponsorship of train travel and accommodation for an invited speaker at a national congress
d. Sponsorship of a trip to Spain for an invited speaker

39. Which of the following statements is MOST likely to be false?

a. The prevalence of a disease is not dependent on disease incidence
b. The prevalence of a disease is dependent on disease duration
c. The incidence of acute disease e.g influenza is often equal to its prevalence
d. The prevalence of hypertension in the typical urban community is likely to exceed the incidence of disease

40. The following statements regarding vitamin D deficiency are correct EXCEPT?

a. Causes rickets in children
b. Incapacitated, bed bound patients are at especially high risk
c. May occur in patients who undergo colectomies of the transverse colon
d. May be caused by rifampicin

41. The National Institute of Clinical Excellence (NICE) has just completed an analysis of 2 intravitreal drugs, X and Y, which are both licensed for treatment of neovascular age related macular degeneration (AMD).
The objective of the analysis was to determine the cost of each drug required to achieve 6/18 visual acuity in the treated eye. The cost of drug X to achieve the target outcome was £550 per treated eye, and Y, £700 per treated eye.
Based on the analysis, NICE recommended that drug X be prescribed for treatment of neovascular AMD. What kind of analysis is described?

a. Cost Utility Analysis
b. Cost Effectiveness Analysis
c. Cost Benefit Analysis
d. Cost Minimisation Analysis

42. Regarding the presenting features of Acute Demyelinating Optic Neuritis (ADON), which of the following statements is MOST likely to be true?

a. Often unilateral in children
b. Disc swelling occurs in 2/3 of patients
c. The commonest field defect is a central or centrocaecal scotoma
d. Disc haemorrhages are rare

43. Which of the following hereditary optic neuropathies is autosomal dominant?

a. Wolfram Syndrome
b. Behr Syndrome
c. Leber's Hereditary Optic Neuropathy (LHON)
d. Kjer Syndrome

44. Which of the following is NOT a feature of Wolfram syndrome?

a. Tall stature
b. Diabetes Insipidus
c. Mental Handicap
d. Optic Atrophy

45. Regarding recurrence rates following initial treatment of previously untreated basal cell carcinomas, which statement is MOST likely to be false?

a. The treatment option with the lowest recurrence rate is Mohs micrographic surgery
b. Cryotherapy carries a recurrence rate of 15% over 5 years
c. The most important predictor of recurrence in a lesion treated with Mohs is the distance of the lesion to the closest resection margin
d. Cumulative recurrence rates for all non - Mohs modalities is in the region of 9%.

46. Regarding the diagnosis, treatment and prognosis of sebaceous gland carcinoma, which of the following statements is MOST likely to be false?

a. Map biopsy of the conjunctiva is important during excision of the primary lesion
b. Alcohol fixation should be avoided
c. Foamy cytoplasm with oil vacuoles is pathognomonic
d. Cure rate with primary radiotherapy is similar to that of Mohs surgery.

47. According to the following scenario, what is the MOST likely diagnosis?

A 6-year-old boy presents with a persistent, right esodeviation.
The angle of the squint is similar (4 prism diopters) at distance and at near. Abduction is limited in his right eye. He has had surgery to correct a squint in the past.

a. Consecutive Esotropia
b. Sensory Esotropia
c. Accommodative Esotropia
d. Cyclic Esotropia

48. The BEST option for treatment of a retinal dialysis without associated retinal detachment is

a. Vitrectomy and silicone oil tamponade
b. Scleral Buckle
c. Laser retinopexy
d. Cryopexy

49. The following peripheral retinal degenerations are associated with an increased risk of rhegmatogenous retinal detachment EXCEPT?

a. Lattice degeneration
b. Retinoschisis
c. Cystic Retinal Tufts
d. Microcystoid degeneration

50. **Which of the following drugs used to treat allergic conjunctivitis does not act as a histamine receptor antagonist?**

a. Ketotifen
b. Olapatadine
c. Lodoxamide
d. Levocabastine

51. **Regarding surgical treatment in Duane's Syndrome, which of the following statements is MOST likely to be false?**

a. Indicated to correct a significant abnormal head posture
b. Most children presenting with Duane's Syndrome have normal stereopsis
c. Surgery should be done as early as possible
d. Medial rectus recession is an option for esotropic Duane's

52. **Pertaining to Ganciclovir, which is MOST likely to be incorrect?**

a. Ganciclovir is administered in its inactive form
b. Ganciclovir is converted to its active form after passing through the liver
c. UL97 mutation renders ganciclovir susceptible to resistance
d. Oral ganciclovir, while not as efficacious as intravenous ganciclovir, carries a lower risk of complications

53. **Which of the following surgical procedures is likely to be the MOST efficacious for the following scenario?**

A 2-year-old girl presents with bilateral, severe ptosis with obstruction of both visual axes. Her eyes are set far apart. Her father has similar features.

a. Aponeurotic repair
b. Frontalis Suspension with Silicone Rods
c. Levator Resection
d. Frontalis Suspension with Fascial Slings

54. **According to NICE guidelines for the management of diabetic retinopathy in pregnancy, which statement is MOST accurate?**

a. Patients with pre-existing diabetes should be screened following their antenatal booking appointment.
b. All patients should be followed up till 6 months after delivery
c. The presence of new vessels is a contraindication to vaginal birth
d. Patients with diabetic retinopathy at first screening should be seen again at 18-20 weeks

55. **The following are situations where sharing of patient information can be done without the patient's explicit consent EXCEPT?**

a. Sharing of information between consultant colleagues involved in the patient's care
b. The use of patient identifiable data for in-house clinical audits
c. Insurance claims
d. Between daycare staff and clinic staff involved in the care of a patient undergoing cataract surgery

56. **According to the following table which tabulates results from a cross sectional study, what are the odds of developing allergic conjunctivitis in a child who has asthma?**

	Allergic Conjunctivitis Present	Allergic Conjunctivitis Absent	Total
Asthma Present	10	4	14
Asthma Absent	5	1	6
	15	5	20

a. 2.5
b. 0.4
c. 2
d. 0.2

57. Which of the following characteristics of Oculopharyngeal Muscular Dystrophy (OPMD) is MOST likely to be true?

a. Onset in childhood
b. Ptosis is an uncommon feature
c. Atrophy of the tongue is often seen
d. Life expectancy is reduced

58. According to NICE guidelines on Hypertension 2011, which of the following anti hypertensive drugs is indicated for patients with essential hypertension who less than 55 years in age?

a. Atenolol
b. Captopril
c. Methyldopa
d. Hydrochlorothiazide

59. The following are risk factors for post occlusion surge EXCEPT?

a. Decreased tubing compliance
b. Decreased bottle height
c. Increased aspiration rate
d. Wound leaks

60. The following statements are TRUE regarding optic disc drusen except?

a. Are usually bilateral
b. Are more obvious early in life
c. Optic discs usually show absent cups
d. Visual field defects and retinal vein occlusion may occur

61. Regarding management of Thygeson's Superficial Punctate Keratitis (TSPK), which of the following statements is MOST likely to be false?

a. Topical ciprofloxacin has been shown to be an effective treatment option
b. Topical idoxuridine is contraindicated
c. Preservative free artificial tears are effective in relieving symptoms
d. Topical corticosteroids are the treatment option of choice

62. The following statements regarding epiblepharon are TRUE except?

a. Most common in Afro-Caribbean children
b. May be associated with dehiscence of lower lid retractors
c. Often resolves with age
d. Surgery involves removal of skin and part of the orbicularis

63. Which of the following statements regarding secreting pituitary adenomas is TRUE?

a. Most frequently present with acromegaly
b. Present initially with bilateral inferotemporal field defects which eventually cross the horizontal meridian
c.. Prolactinomas frequently require surgery
d. Octreotide is frequently ineffective as monotherapy for growth hormone secreting tumors

64. The following statements regarding orbital varices are TRUE except?

a. Are a differential diagnosis for patients with intermittent proptosis
b. Are commonly associated with intracranial extension
c. May cause orbital haemorrhage
d. Optic neuropathy caused by surgery is usually due to vascular compromise

65. Which of the following statements is LEAST likely to be true regarding Congenital Rubella infection?

a. is a single stranded RNA virus
b. Risk of congenital defects is highest if exposure occurs in the first trimester
c. The most common systemic manifestations are cardiac defects
d. The classic 'salt and pepper' retinopathy is associated with a normal ERG

66. Which of the following histopathologic appearances describes cribiform adenocystic carcinoma of the lacrimal gland?

a. Nuclear palisading with nests and cords of tumour cells
b. Basophilic, lobulated appearance with pools of mucin
c. Eosinophilic cytoplasm with cross striations
d. Lobulated appearance due to lipid globules which stain with Oil Red O

67. Regarding investigations and management of Brown's syndrome, which of the following statements is MOST likely to be false?

a. In non-traumatic, acquired Browns, rheumatoid arthritis should be excluded
b. MRI of the orbit may be helpful in cases of congenital Brown's
c. Congenital Brown's rarely resolves spontaneously
d. Indications for surgery include a debilitating abnormal head posture

68. For the following scenario, which is the MOST likely diagnosis?

The Paediatric SHO who detected leukocoria refers a 5-month-old baby girl with linear truncal hyperpigmentation to your clinic. Fundal examination, reveals bilateral tractional retinal detachments.

a. Incontinentia Pigmenti
b. Norrie's Disease
c. Walker Warburg
d. Xeroderma Pigmentosa

69. Which of the following statements regarding Mobius syndrome is MOST likely to be false?

a. Is sporadic in inheritance
b. An exotropia is often present
c. Exposure keratopathy is a feature
d. Forced duction tests are often positive

70. The normal, base-in fusional reserves for distance are

a. 5-10 Prism Dioptres (PD)
b. 15-20 PD
c. 25-30 PD
d. 35-40 PD

71. Regarding solar retinopathy, which of the following statements is MOST likely to be false?

a. Often associated with subsequent full thickness macular hole
b. Presentation is often within days of damage occurring
c. Damage most often occurs in the dominant eye
d. Most common symptoms include a central scotoma and negative afterimage of the sun

72. Regarding treatment of macular holes, which of the following statements is MOST likely to be true?

a. Vitrectomy is useful in prophylactically preventing formation of full thickness macular holes
b. Treatment of full thickness macular holes with laser retinopexy is a viable option
c. Regarding vitrectomy for full thickness macular holes, eyes with worse pre-operative visual acuity improve to a greater degree than eyes with better pre-operative acuity
d. The use of autologous platelets improves final post-vitrectomy visual acuity

73. The following are findings in chronic rhegmatogenous retinal detachment EXCEPT?

a. Absolute visual field defect
b. Demarcation line
c. Retinal Cysts
d. Neovascular glaucoma

74. Regarding diagnosis and management of Herpes Zoster Ophthalmicus (HZO), which of the following statements is MOST likely to be false?

a. The Tzanck smear is cheap and easy to perform
b. The Tzanck smear is highly sensitive
c. Direct fluorescent antibody (DFA) testing is the investigation of choice when acute diagnosis is required
d. DFA testing has a lower false negative rate when compared to the Tzanck smear

75. Which of the following statements is MOST likely to be true about phacoanaphylactic uveitic glaucoma?

a. Is a non- granulomatous inflammatory glaucoma
b. There is a significant Type 1 hypersensitivity response
c. Inflammation usually occurs within an hour of capsular disruption during trauma
d. Histoogically, there is zonal inflammation with polymorphonuclear leucocytes (PMNL) surrounding lens fragments, in turn surrounded by macrophages and giant cells

76. The following statements regarding aqueous misdirection syndrome are likely to be TRUE except?

a. Typically occurs in patients with open angle glaucoma who have undergone filtration surgery
b. Patients present with a flat anterior chamber
c. Discontinuation of cycloplegics may precipitate aqueous misdirection
d. The anterior chamber typically remains shallow despite a patent iridotomy

77. Corneal guttata are seen in the following conditions EXCEPT?

a. Fuch's Hereditary Endothelial Dystrophy (FHED)
b. Congenital Hereditary Stromal Dystrophy (CHSD)
c. Iridocorneal Endothelial Syndrome (ICE)
d. Posterior Polymorphous Dystrophy (PPMD)

78. The following statements regarding conjunctival melanoma are TRUE except?

a. Is an uncommon lesion
b. Cases arising from primary acquired melanosis are less common as compared to those arising *de novo*
c. Haematogenous metastasis is well documented
d. Direct extension into the orbit is possible

79. The following are likely to cause synechial angle closure glaucoma EXCEPT?

a. Proliferative Diabetic Retinopathy
b. Iridocorneal Endothelial Syndrome
c. Silicone Oil induced glaucoma in a phakic eye
d. Posterior Polymorphous Dystrophy

80. Which of the following statements regarding Primary Acquired Melanosis (PAM) MOST likely to be false?

a. is usually seen in middle aged or elderly white patients
b. Is acquired, usually with a history of inciting trauma or inflammation
c. Is most often on the bulbar conjunctiva
d. Nodularity is a sign of malignant change

81. What is the inheritance pattern of gyrate atrophy?

a. Autosomal Recessive
b. Mitochondrial
c. X Linked
d. Autosomal Dominant

82. What proportion of patients with angioid streaks develop associated choroidal neovascularization?

a. 10%
b. 25%
c. 50%
d. 75%

83. In which of the following genes are mutations responsible for Meesman's dystrophy?

a. K3
b. BIGH3
c. Gelsolin
d. K5

84. Regarding radiation retinopathy (RR), what is the minimum dose of cephalic radiation known to put the patient at risk?

a. 15 Gy
b. 30 Gy
c. 50 Gy
d. 70 Gy

85. The following statements are true regarding the optic radiations EXCEPT?

a. May be damaged within the temporal lobe
b. Form the most anterior part of the internal capsule
c. When damaged within the internal capsule, may be associated with ocular deviation towards the side of the lesion
d. Unilateral lesions do not affect visual acuity

86. Which of the following features is LEAST suggestive of a diagnosis of keratoconus?

a. Regular astigmatism
b. Inferior corneal steepening
c. Keratometric values of >47D
d. Multiple inadequate spectacle corrections

87. Which of the following is the MOST likely diagnosis for the following scenario?

A 20-year-old gentleman presents with unilateral blurring of vision. On examination he has a right fundal orange retinal lesion involving the macula with exudation. There appears to be a feeding artery and draining vein. Fluorescein angiography shows rapid filling of the feeding artery and subsequently, the lesion, accompanied by late diffuse leakage.

a. Retinal Capillary Haemangioma
b. Racemose Haemangioma
c. Retinal Cavernous Haemangioma
d. Retinal Macroaneurysm

88. Regarding Bell's palsy, which of the following features is MOST likely to be false?

a. Is the commonest cause of unilateral lower motor neurone (LMN) 7th nerve palsy
b. Is strongly linked with a viral etiology
c. Spontaneous recovery is rare
d. Steroids increase the chances of recovery

89. Paradoxical pupillary responses are seen in the following conditions EXCEPT?

a. Oguchi's Disease
b. Blue Cone Monochromatism
b. Fundus albipunctatus
c. Rod Monochromatism

90. In a randomised controlled trial, data may go missing due to patient drop out. This may cause some degree of bias.
The following are methods that are introduced to minimize said bias EXCEPT?

a. Worse case scenario analysis
b. Intention-to-treat analysis
c. Last Observation Carried Forward
d. Treatment allocation concealment

Paper 1 Answers and Discussion

1. Answer: c

Discussion: Of the listed conditions, only Knobloch syndrome is autosomal recessive.

2. Answer: c

Discussion: Vancomycin, Teicoplanin and Gentamicin are effective against MRSA and may be used for deep-seated infection. De-colonization is usually achieved with the use of mupirocin nasal ointment and antiseptic body wash for up to 5 days. Fluoroquinolone antibiotics should be avoided as treatment options for MRSA infection.

Source:
1. Deresinski S. Vancomycin in Combination with Other Antibiotics for the Treatment of Serious Methicillin-Resistant *Staphylococcus aureus* Infections. *Clinical Infectious Diseases* Volume 49, Issue 7 Pp. 1072-1079

3. Answer: c

Discussion: Oblique astigmatism is a form of regular astigmatism where the meridians of greatest and least power do not lie at 90 or 180 degrees.
Against-the-rule astigmatism is regular astigmatism with its steepest meridian lying at 180 degrees.
Astigmatism in keratoconus is often irregular, in that the meridians of greatest and least power are not perpendicular.
Children often have with-the-rule astigmatism and adults tend to have against-the-rule astigmatism.

4. Answer: d

Discussion: The clotting cascade consists of two parallel pathways that both end in the conversion of pro-thrombin to thrombin. The intrinsic pathway requires contact activation with sub-endothelial or extravascular collagen and is thought to be the predominant pathway involved in the inflammatory response.
The extrinsic pathway is triggered by endothelial damage, exposing expressed tissue factor (Factor 3) to circulating Factor 7.
Relevant co-factors include vitamin K (required for clotting factor production by the liver) and calcium. (required for activation of circulating clotting factors) and Calcium. Vitamin C is not a relevant co-factor.

5. Answer: a

Discussion: Tuberous sclerosis is the second most common phakomatosis or neurocutaneous disorder. With a prevalence of 1/6000, it is second only to Neurofibromatosis Type 1 with a prevalence of 1/4000. Neurofibromatosis Type 2 is a distant third, with a prevalence of 1/40000.
Von Hippel Lindau and Sturge Weber are rare.

Source:
1. Borkowska J, Schwartz RA, Jozwiak S. Recent perspectives on diagnosis and treatment of tuberous sclerosis complex in children. *Int J Disability Human Development*. 2009;8:369-375.
2. Denniston AKO, Murray PI. 2006. *Oxford Handbook of Ophthalmology*. Oxford. Oxford University Press

6. Answer: c

Discussion: Ophthalmologically significant DNA viruses include **H**erpes Simplex, **V**aricella Zoster, **C**ytomegalovirus, **A**denovirus and **M**olluscum contagiosum. (mnemonic: **D**id you know, **H**is **V**accine **C**ured **A**dam and **M**ary?) all of which are double stranded.

Rubella is a single stranded RNA virus of the *Togaviridae* family.

7. Answer: b

Discussion: Herpes Zoster does not cause saddle nose, which is caused by inflammation and subsequent destruction of nasal cartilage.

8. Answer: c

Discussion: All the phakomatoses are autosomal dominant except Wyburn Mason Syndrome and Sturge Weber (which are both sporadic) as well as Ataxia Telangiectasia (recessive).
Ataxia Telangiectasia (also known as Louis Bar Syndrome) presents at the age of 2 years with ataxia, conjunctival telangiectasia, limitation of ocular versions and nystagmus.

Source:
1. Taylor AMR, Byrd PJ. Molecular Pathology of Ataxia Telangiectasia. *J Clin Pathol* 2005;58:1009-1015

9. Answer: b

Discussion: Posterior polar cataracts are associated with remnants of the vascular hyaloid system. They are typically inherited as a dominant trait, but may be sporadic. The genetic mutation has been mapped to locus 16q22. The cataracts present as oval rosettes and are associated with an increased risk of posterior capsule rupture during surgery.

Source:
1. Maumenee III. Classification of hereditary cataracts in children by linkage analysis. *Ophthalmology* 1979;86:1554–8

10. Answer: c

Discussion: Risk factors for conversion of ocular hypertension to POAG which were demonstrated in the ocular hypertension study (OHTS) include higher IOP, larger cup disc ratio, greater pattern standard deviation (PSD), older age, and thinner central corneal thickness.

Source:
1. Kass MA et al. The Ocular Hypertension Treatment Study: a randomized trial determines that topical ocular hypotensive medication delays or prevents the onset of primary open-angle glaucoma. *Arch Ophthalmol.* 2002 Jun;120(6):701-13; discussion 829-30.

11. Answer: d

Discussion: Final visual acuity in CRVO largely depends on the presenting acuity.
1/3 of patients with ischaemic CRVO develop neovascular glaucoma within 4 months of the onset.
20% of patients presenting with an initial acuity of 6/60 or better improve spontaneously to 6/15 or better, whereas 80% of those presenting with an acuity of worse than 6/60 do not improve or may worsen.

Source:
1. Royal College Interim Guidelines for Retinal Vein Occlusion 2010

12. Answer: d

Discussion: Kollner's rule states that macular disorders give rise to blue – yellow colour vision defects and optic neuropathies give rise to red – green colour defects. There are notable exceptions to the rule, for example glaucoma, acute demyelinating optic neuritis (which may present with fluctuating colour vision defects) and cone dystrophies.

13. Answer: b

Discussion: The differentials for cutaneous keratotic lesions include premalignant lesions such as keratoacanthomas and actinic keratosis as well as malignant lesions such as squamous cell carcinoma and basal cell carcinoma.
Ulceration in association with a keratotic lesion suggest a diagnosis of squamous cell carcinoma. Keratin pearls are desmosomes in association with intracellular keratin filaments and are pathognomonic of squamous cell carcinoma.

14. Answer: d

Discussion: The commonest causes of orbital cellulitis in children are Streptococcus pneumoniae and Haemophilus influenzae which are almost invariably associated with sinusitis. Ethmoidal sinusitis is the commonest cause of orbital cellulitis at all ages, especially in very young children who have not yet developed their frontal sinuses. The incidence of orbital cellulitis tends to peak during the winter months secondary to an increase in the number of viral upper respiratory tract infections, and shows a male preponderance

Source:
1. Harrington JN. *Orbital Cellulitis.* Emedicine.medscape.com

15. Answer: b

Discussion: Classification of esodeviations can be divided into whether they are *concomitant* or *incomitant*. Concomitant deviations can further be divided into whether they are *constant* or *intermittent*.
The above scenario describes a concomitant deviation with cross fixation which has been present since birth. The most likely diagnosis is therefore a congenital esotropia.

16. Answer: c

Discussion: According to the United States Eye Injury Registry, 51% of patients with traumatic hyphaema have associated posterior segment injury. This indicates the importance of imaging of the posterior segment despite a poor fundal view. Elevated IOP occurs in up to a quarter, and corneal bloodstaining may occur despite a normal IOP. A black ball hyphaema indicates deoxygenation of the blood, and increases hypoxic risk to the endothelium, increasing the risk of corneal bloodstaining.

Source:
1. Beyer TL, Hirst LW (1985) Corneal blood staining at low pressures. *Arch Ophthalmol* 103: 654–655
2. Ferenc Kuhn. 2008. *Ocular Traumatology.* Springer.

17. Answer: c

Discussion: Of the above conditions, Marfan's syndrome does not present with an optically empty vitreous. An optically empty vitreous is defined as 'lacunae within early onset syneretic vitreous'.
It is best observed through a widely dilated pupil.

18. Answer: a

Discussion: *Neisseria gonorrhoea* is a cause of ophthalmia neonatorum and typically presents within the first 24-48 hours of life. The diplococci grow well on chocolate agar incubated with carbon dioxide, and may be isolated on Thayer Martin agar. Nutrient agar cultures *Acanthamoeba.* MacConkey is used to culture enteric bacteria.

19. Answer: c

Discussion: This scenario points to functional stenosis of the nasolacrimal passage. Lacrimal scintigraphy is probably the best investigation as the test examines the tear drainage under physiological conditions.

20. Answer: c

Discussion: While cerebral angiography remains the gold standard for detection of intracranial aneurysms, CT angiograms are highly sensitive and are non invasive. The logical step would be to proceed with a CT angiogram, and if negative, cerebral DSA should then be considered.

Source:
1. Teksam M. et al. Multi-Section CT Angiography for Detection of Cerebral Aneurysms. *AJNR* 2004 25: 1485-1492

21. Answer: d

Discussion: Goldman tonometry is based on the invalid assumption that there is little variability between human eyes in terms of corneal thickness and rigidity. Readings may be falsely low in patients with corneal oedema or thin corneas, leading to a diagnosis of normotension glaucoma in patients who may actually have chronic open angle glaucoma with high pressure. Readings may be artificially high in patients with thick or scarred corneas.

22. Answer: b

Discussion: The role of contrast sensitivity testing in clinical practice remains controversial. Detection of reduced visual function in various disorders like cataract, glaucoma and cerebral lesions may preclude deterioration in acuity. However, patterns of loss are not disease specific, reducing diagnostic value. Contrast sensitivity remains an important parameter of daily visual function, and has been measured as a secondary outcome in studies such as the ONTT and the TAP study.

Source:
1. David Hinton, SriniVas Sadda, Andrew P. Schachat, Charles P. Wilkinson, Peter Wiedemann. 2012. *Retina*. 5th Ed. Elsevier Health Sciences.

23. Answer: a

Discussion: This is likely to be Wegener's Granulomatosis, given the typical triad of orbital involvement, pulmonary, and renal involvement. C - ANCA antibodies may be detectable in close to 100% of patients using both immunofluorescence and ELISA.

Source:
1. Moosig F, Lamprehct P, Gross WL. Wegener's Granulomatosis: The Current View. *Clin Rev Allergy Immunol*. October 2008;35(1-2):19-21

24. Answer: b

Discussion: While the clinical appearance may suggest either a melanoma, metastasis or a haemangioma, melanomas have low internal reflectivity and may demonstrate dual circulation on fluorescein angiography, whilst haemangiomas demonstrate early hyperfluorescence on angiography due to rapid vascular filling.

25. Answer: a

Discussion: The genetic makeup of uveal melanoma is increasingly being used to predict mortality. Monosomy 3 is associated with loss of the gene coding for BAP1 and is predictive of metastatic disease. Gains of chromosome 8q and 6p are associated with a better and worse prognosis respectively.

Source:
1. Sandinha MT, Farquharson MA, McKay IC, Roberts F. Monosomy 3 Predicts Death but Not Time until Death in Choroidal Melanoma. *Investigative Ophthalmology & Visual Science*, October 2005, Vol. 46, No. 10
2. www.ocularmelanoma.org

26. Answer: c

Discussion: The structure of the ocular layers presents considerable barriers to penetration of drug molecules.
Normal basal tear secretion significantly dilutes the concentration of instilled drug and hastens its removal.
Tight junctions between corneal epithelial cells impedes movement of hydrophilic molecules.
The converse is true for the water rich corneal stroma and sclera.

27. Answer: c

Discussion: The inflammation in PIC usually subsides with the formation of atrophic yellow-white scars which predispose to secondary CNV formation.

Source:
1. Watzke RC, Packer AJ, Folk JC, et al. Punctate inner choroidopathy. *Am J Ophthalmol*. 1984;98:572-584.

28. Answer: b

Discussion: A repeat PK is likely to fail given the harsh microenvironment of the patient's eyes. An OOKP, which is particularly resilient in patients with chronic dry eye, would be the option of choice to restore her vision.

Source:
1. Liu C, et al. The Osteo-Odonto-Keratoprosthesis. *Semin Ophthalmol*. 2005 Apr-Jun;20(2):113-28.

29. Answer: a

Discussion: indications for SLT include POAG, PDS, PEX, and primary angle closure glaucoma (PACG) without synechial closure. Contraindications include ICE syndrome, neovascular and uveitic glaucoma. The lack of response in one eye is a relative contraindication to treatment of the fellow eye.

30. Answer: c

Discussion: TASS is an acute, non infectious inflammation of the anterior chamber seen after intraocular surgery. Various aetiologies include contaminated irrigating fluids, preserved medications, heat stable bacterial toxins remaining on surgical instruments after autoclaving, topical ointments and surgical glove talc. Patients present within 12-24 hours post surgery (in contrast to the 72-96 hours for infectious endophthalmitis) with mild discomfort, generalised corneal oedema and intense, fibrinous inflammation. In contrast to endophthalmitis, TASS responds rapidly to intensive topical steroids.

Source:
1. Al-Ghoul AR. Toxic Anterior Segment Syndrome. *emedicine.medscape.com*
2. Holland SP, Morck DW, Lee TL. Update on Toxic Anterior Segment Syndrome. *Curr Opin Ophthalmol.* 2007 Feb;18(1):4-8.

31. Answer: c

Discussion: While the diagnosis of SJS is clinical, and there are no specific tests which allow a definite diagnosis, biopsies from inflamed conjunctiva, together with immunohistochemical staining, may reveal sub-epithelial plasma cell and lymphocyte infiltration.
Antibody and complement deposition within vessel walls and epithelium basement membrane is also typically seen. Cessation of the offending drug or treatment of the offending infection is currently the backbone of management of active disease.

Systemic corticosteroid therapy is still controversial. A recent multicenter European trial revealed no benefit in terms of survival rates in patients administered high dose corticosteroids early in the disease. The aims of ophthalmic management are threefold; **suppress ocular inflammation, aid re-epithelisation, and manage long term complications,** including any subsequent corneal scarring. Lamellar keratoplasties should only be considered after disease quiescence of 3-6 months.

Source:
1. Sekula P, Dunant A, Mockenhaupt M, Naldi L, Bouwes Bavinck JN, Halevy S, et al. RegiSCAR study group. *Comprehensive survival analysis of a cohort of patients with Stevens-Johnson syndrome and toxic epidermal necrolysis.* J Invest Dermatol. 2013 May;133(5):1197-204.

32. Answer: a

Discussion: Contact dermatitis of the periocular region is highly suggestive of drop hypersensitivity. Inferior tarsal/fornical follicles are suggestive of a drop hypersensitivity reaction, but may also occur in both adenoviral and chlamydial conjunctivitis.

Source:
1. American Academy of Ophthalmology Corneal/External Disease Panel. Preferred Practice Pattern: Conjunctivitis. San Francisco, Ca: AAO; 2003.

33. Answer: d

Discussion: Steroid induced glaucoma is a form of open angle glaucoma. There is no gender bias. Topically instilled steroids as well as intravitreal steroids carry the highest risk followed by, in order of decreasing risk, intravenous, parenteral and inhaled steroids. 90% of patients with primary open angle glaucoma (POAG) demonstrate some form of steroid response as compared to 30% of patients with normal eyes. Eyes with PAC behave similarly to normal eyes in terms of steroid response.

Source:
1. Armaly MF. Effect of corticosteroids on intraocular pressure and fluid dynamics. I. The effect of dexamethasone in the normal eye. *Arch Ophthalmol.* 1963;70:482.
2. Armaly MF. Effect of corticosteroids on intraocular pressure and fluid dynamics. II. The effect of dexamethasone in the glaucomatous eye. *Arch Ophthalmol.* 1963;70:492.

34. Answer: b

Discussion: The differential diagnoses would include choroidal neovascularisation in the setting of myopia or previous trauma. However, the location and orientation of the scar is typical of choroidal rupture and thus, is the likeliest diagnosis.

Source:
1. Kuhn F. 2008. *Ocular Traumatology.* Springer.

35. Answer: d

Discussion: Of the above, the least likely life threatening complication is raised intracranial pressure. While it has been reported to occur in association with retinitis pigmentosa, it is uncommon. Cardiac conduction defects are associated with Kearn Sayre syndrome, renal impairment in both Alport's and Bardet Biedl, which may cause mortality in both. Liver cirrhosis occurs in association with Bassen Kornzweig.

Source:
1. medscape.emedicine.com
2. Donin JF, Crowley LG. Papilledema complicating retinitis pigmentosa. *AMA Arch Ophthalmol.*1958 Apr;59(4):609-11.

36. Answer: d

Discussion: Irvine Gass syndrome refers to post operative (cataract) cystoid macular oedema (CMO). Risk factors include previous uveitis, diabetes, capsular rupture and/or vitreous loss. While topical prostaglandins have been associated with IGS, there is no clear evidence that drug cessation pre operatively reduces risk. Patients typically complain of central blurring of vision, a reduced foveal light reflex may be seen, and fluorescein angiography reveals typical petalloid leakage. The majority of patients attain 6/9 vision or better after a year.

Source:
1. Panteleontidis V, Detorakis ET, Pallikaris IG, Tsilimbaris MK. Latanoprost-Dependent Cystoid Macular Edema Following Uncomplicated Cataract Surgery in Pseudoexfoliative Eyes. *Ophthalmic Surg Lasers Imaging.* Mar 9 2010;1-5.
□□□Law SK, Kim E, Yu F, Caprioli J. Clinical cystoid macular edema after cataract surgery in glaucoma patients. *J Glaucoma.* Feb 2010;19(2):100-4.

37. Answer: b

Discussion: According to 2013 Royal College Guidelines on Age Related Macular Degeneration, contraindications to intravitreal ranibizumab include a visual acuity of less than 6/96, the presence of permanent structural damage, a lesion size of 12 disc diameters or more, and documented hypersensitivity to ranibizumab.

Source:
1. Royal College of Ophthalmology Guidelines on Age Related Macular Degeneration 2013

38. Answer: b

Discussion: According to Ethical Standards for Commercial Sponsorship for the NHS, the following need to be declared.

'NHS funding from an external source, including funding of all or part of the costs of a member of staff, NHS research, staff, training, pharmaceuticals, equipment, meeting rooms, costs associated with meetings, meals, gifts, hospitality, hotel and transport costs (including trips abroad), provision of free services (speakers), buildings or premises.'

'Personal gifts costing less than £25 per gift do not need to be declared, however, multiple small gifts from the same source or closely related sources amounting to £100 or more in the same calendar year should be declared.'

Source:
1. Department of Health. *Commercial Sponsorship – Ethical Standards for the NHS.* Nov 2000

39. Answer: a

Discussion: The prevalence of a disease is the total number of cases in a sample population at any given time, and is directly proportional to the incidence (the number of new cases in a sample population per unit time) and duration of the disease. The prevalence of chronic disease exceeds incidence, while prevalence often is equal to incidence in the case of acute disease of short duration.

40. Answer: c

Discussion: Vitamin D deficiency may be caused by poor nutrition, lack of sunlight (patients who are bed bound and institutionalised are particularly at higher risk), malabsorption due to celiac spue and disorders affecting the small intestine including cystic fibrosis and medications that induce hepatic p450 enzymes including rifampicin. It causes rickets in children and osteomalacia in adults.

Source:
1. Liu BA, Gordon M, Labranche JM et al. Seasonal prevalence of vitamin D deficiency in institutionalised older adults. *J Am Geriatr Soc.* May 1997;45(5):598-603
2. Tangpricha V, Luo M, Fernandez-Estivariz C, et al. Growth hormone favorably affects bone turnover and bone mineral density in patients with short bowel syndrome undergoing intestinal rehabilitation. *J Parenter Enteral Nutr.* Nov-Dec 2006;30:480-6

41. Answer: b

Discussion: Cost Effectiveness Analysis (CEA) involves calculation of the cost of alternative treatment options needed to achieve a target outcome – in this case, 6/18 visual acuity in the treated eye. The most affordable option is usually the one selected.

Source:
1. Gray AM, Clarke PM, Wolstenhome JL, Wordsworth S. 2011. *Applied Methods of Cost Effectiveness Analysis in Healthcare.* New York. Oxford University Press.

42. Answer: c

Discussion: Episodes of ADON are usually unilateral in adults and bilateral in children, and are usually the initial demyelinating event in patients who are eventually diagnosed with multiple sclerosis. Patients present with sudden, painful blurring of vision as well as pain on eye movement. Visual acuity is variable on presentation, with variable colour vision defects. A relative afferent pupillary defect is usually present. Disc swelling is present in 1/3 of patients. Haemorrhages are rare.

43. Answer: d

Discussion: LHON has a mitochondrial inheritance. Both Behr and Wolfram syndromes are recessive, while Kjer syndrome is autosomal dominant.

44. Answer: a

Discussion: Wolfram syndrome is characterized by hereditary, early onset (1st decade) optic neuropathy associated with short stature, mental handicap, diabetes mellitus, diabetes insipidus and anosmia.

45. Answer: b

Discussion: Recurrence rates are lowest with Mohs surgery, around 1%. The most important predictor of recurrence is the distance from the lesion to the nearest resection margin.
Non-Mohs modalities carry a cumulative recurrence rate of 9%, with the highest being surgical excision – carrying a recurrence rate of 10%. Cryotherapy and radiation carry a recurrence rate of 7.5% and 8.7% respectively.

Source:
1. Bader RS. *Basal Cell Carcinoma Treatment and Management.* emedicine.medscape.com

46. Answer: d

Discussion: Mohs micrographic surgery is the best mode of treatment for sebaceous gland carcinomas. Map biopsies of the conjunctiva should be obtained at the same time as excision of the primary lesion as pagetoid, multicentric spread tends to occur throughout the lid to involve the conjunctiva. Specimens should be formalin fixed and frozen and stained to identify fat vacuoles, typically with oil red o stain. The histopathology staff should be informed to avoid alcohol fixation. Radiotherapy should not be a first option for treatment because of high rates of complications and recurrence as well as non - availability of a histopathologic diagnosis.

Source:
1. Wali UK, Al-Mujaini A. Sebaceous gland carcinoma of the eyelids. Oman J Ophthalmol. 2010 Sep-Dec; 3(3): 117–121

47. Answer: a

Discussion: Consecutive esotropias usually arise as a result of surgery to correct an exodeviation. They are small angled deviations, measuring less than 10 prism diopters.

48. Answer: c

Discussion: Retinal dialyses are caused by tearing of the retina at its attachment to the ora. Vitreous traction on the torn retina remains. Traumatic retinal dialyses tend to present early but progression to a retinal detachment is usually delayed. Laser retinopexy is the treatment of choice for retinal dialyses without associated retinal detachment.

Source:
1. Kennedy C, Parker C, McAllister I (1997) Retinal detachment caused by retinal dialysis. *Aust N Z J Ophthalmol* 251: 25–30

49. Answer: d

Discussion: Of the above, microcystoid degeneration does not predispose to retinal detachment

Source:
1. Lewis H. *Peripheral retinal degenerations and the risk of retinal detachment.* Am J Ophthalmol. 2003 Jul;136(1):155-60.

50. Answer: c

Discussion: Drugs used to treat allergic conjunctivitis fall into a few categories: *Mast Cell Stabilisers, Anti Histamines* and *Corticosteroids.*
Ketotifen, Olapatadine and Lodoxamide are mast cell stabilisers. However, the latter lacks any histamine receptor antagonism. Levocabastine is a H1 selective receptor antagonist

Source:
1. medscape.emedicine.com

51. Answer: c

Discussion: The majority of children with Duane's Syndrome have stereopsis, and as surgery is often disappointing and does nothing to improve fusional ability, it should only be considered in the setting of a debilitating abnormal head posture, and performed as late as possible, for fear of disrupting the development of normal binocular function.

Source:
1. Verma A. Duane Syndrome. *Emedicine.medscape.com*
2. Kraft SP. A surgical approach for Duane syndrome. *J Pediatr Ophthalmol Strabismus.* May-Jun 1988;25(3):119-30

52. Answer: b

Discussion: Ganciclovir may be administered intravenously, orally, or intravitreal via injections or implants. It is activated via phosphorylation after uptake by virus-infected cells. The UL97 mutation renders ganciclovir susceptible to resistance. Ganciclovir may be administered for induction of treatment or maintenance therapy, at least till CD4 counts recover on HAART.
Oral ganciclovir is not as efficacious as intravenous ganciclovir but it carries a lower risk of complications such as neutropenia.

Source:
1. emedicine.medscape.com

53. Answer: b

Discussion: Ptosis in the setting of blepharophimosis is typically managed with frontalis suspension. Traditionally, in a child below the age of 3, the amount of fascia lata available is usually insufficient and therefore synthetic slings are the preferred material. However, recent literature suggests that autogenous fascia lata may be used successfully in children under 3 years of age.

Source:
1. Crawford JS. Repair of ptosis using frontalis muscle and fascia lata. *Trans Am Acad Ophthalmol Otolaryngol.* 1956 Sep-Oct; 60(5):672-8.
2. Leibovitch I, Leibovitch L, Dray JP. Long-term results of frontalis suspension using autogenous fascia lata for congenital ptosis in children under 3 years of age. Am J Ophthalmol. 2003;136:866–71.

54. Answer: a

Discussion: Patients with pre existing diabetes should be screened following their antenatal booking appointment, unless they have not had a fundal examination within the past year, whereby they should be screened immediately.
They should then have a follow-up at 28 weeks, unless diabetic retinopathy was present at their initial screening, whereby they should be seen again at 18-20 weeks.
Diabetic retinopathy is not a contraindication to vaginal delivery.

Source:
1. National Institute of Clinical Excellence Guideline CG63: Diabetes in Pregnancy

55. Answer: c

Discussion: Situations in which sharing of identifiable patient information can be done with the patient's implied consent include information sharing between clinical staff involved in patient care, as well as in-house clinical audits – as long as the patient has access to information regarding said audits and has not objected.

Source:
1. General Medical Council Guidelines on Confidentiality October 2009

56. Answer: a

Discussion: Odds within a group are calculated as the **number with disease/number without disease**, which in this case would be **10/4 = 2.5**

57. Answer: c

Discussion: OPMD is an autosomal dominantly inherited disorder characterised by late onset ptosis (mean age of onset reported to be 48.1 years), progressive dysphagia (mean onset 50.7 years) and muscular weakness (tongue atrophy – seen in 82%, lower limb weakness –seen in 71%, and limitation of upgaze – seen in 62%). The majority of individuals eventually need a wheelchair, but life expectancy is not reduced.

Source
1. Bouchard JP, Brais B, Brunet D, Gould PV, Rouleau GA. Recent studies on oculopharyngeal muscular dystrophy in Quebec. *Neuromuscul Disord.* 1997;7:S22–9

58. Answer: b

Discussion: As per NICE guidelines on Hypertension 2011, patients below the age of 55 with essential hypertension should be prescribed either an ACE inhibitor or angiotensin receptor blocker as first line treatment.

Source:
1. National Institute of Clinical Excellence Guideline CG127: Hypertension

59. Answer: a

Discussion: Post occlusion surge refers to the sudden shallowing of the anterior chamber that tends to occur during the aspiration of particularly dense nuclear fragments.
With continuous aspiration, negative pressure builds up within the phaco system as the phaco tip is occluded, leading to sudden aspiration of a large amount of fluid as the occlusion clears.
This has the effect of suddenly causing the anterior chamber to shallow, increasing the risk of posterior capsular rupture. This tends to occur in the setting of large, mature cataracts that are often associated with floppy posterior capsules.
Risk factors include increased tubing compliance, decreased bottle height, increased aspiration rate and wound leaks.

Source:
1. Ward MS, Georgescu D, Olson RJ. Effect of bottle height and aspiration rate on postocclusion surge in Infiniti and Millennium peristaltic phacoemulsification machines. *J Cataract Refract Surg.* 2008 Aug;34(8):1400-2

60. Answer: b

Discussion: Optic disc drusen are usually bilateral and become progressively more obvious throughout life. The discs are lumpy, with absent cups. Optociliary shunts may occur. Sight threatening complications include visual field defects and retinal vein occlusion.

61. Answer: a

Discussion: The mainstay of management includes topical corticosteroids e.g fluorometholone 0.1% tapered gradually over a few months. Extended wear bandage lenses, preservative free artificial tears, and topical ciclosporin have all been proven effective in relieving symptoms and clinical signs. Topical antibiotics have not been shown to be effective, and whilst there have been mixed reviews regarding efficacy of topical antivirals, idoxuridine causes scarring and is thus contraindicated.

Source:
1. Tabbara KF, Ostler HB, Dawson C, Oh K. Thyegeson's Superficial Punctate Keratitis. *Ophthalmology.* Jan 1981;88(1)75-7
2. Braley AE, Alexander RC. Superficial punctate keratitis; isolation of a virus. *AMA Arch Ophthalmol.* Aug 1953;50(2):147-54
3. Reinhard T, Sundmacher R. Topical cyclosporine A in Thygeson's Superficial Punctate Keratitis. *Graefes Arch Clin Exp Ophthalmol.* Feb 1999;237(2):109-12

62. Answer: a

Discussion: Epiblepharon is an often bilateral, horizontal fold of skin running parallel to the lower lid margin. It is associated with dehiscence of the lower lid retractors, and pre-tarsal orbicularis override by the pre-septal orbicularis. This causes pseudo-trichiasis and corneal irritation. It often improves with age. Surgery is indicated for symptomatic corneal irritation, and involves removal of skin and part of the orbicularis, concluding with attachment of the upper edge of the skin incision to the tarsal plate.

Source:
1. Gladstone GJ, Black EH, Myint S, Brazzo BJ, Nesi FA. 2002. *Oculoplastic Surgery Atlas: Eyelid Disorders.* New York. Springer-Verlag.

63. Answer: d

Discussion: Pituitary adenomas may be divided into micro (<10 mm) or macro (≥10 mm) adenomas. Most pituitary adenomas are microadenomas. Macroadenomas tend to present with pressure effects, (classically with bilateral superotemporal field defects) but may also be secretory, presenting with hormonal imbalance. The commonest secretory adenomas are prolactinomas which are notoriously sensitive to dopaminergic agonists and may be treated with cabergoline or bromocriptine.
Growth hormone secreting tumours are aggressive and generally require oral octreotide as a form of temporary chemoreduction while awaiting definitive therapy which includes surgery and radiotherapy.

Source:
1. Biller BM, Molitch ME, Vance ML, Cannistraro KB, Davis KR, Simons JA, et al. Treatment of prolactin-secreting macroadenomas with the once-weekly dopamine agonist cabergoline. *J Clin Endocrinol Metab.* Jun 1996;81(6):2338-43.
2. Diez JJ, Iglesias P. Current management of acromegaly. *Expert Opin Pharmacother.* Jul 2000;1(5):991-1006.

64. Answer: b

Discussion: Orbital varices are rare, vascular lesions that intermingle intimately with orbital tissues. Distensibility of the vessel walls is a typical feature, allowing engorgement caused by posture, straining or coughing – leading to the typical presentation of intermittent proptosis.
Rarely, abnormalities of the bony orbit may occur (reported to be as low as 4% in a large case series from Moorfields), with lesions reported to involve cerebral parenchyma through defects in the orbital roof.
The lesions may be complicated with orbital haemorrhage, and surgery is indicated in complicated lesions. Surgery is fraught with danger, as the lesions are friable and bleed easily. Optic nerve compromise may occur, usually due to ischaemia.

Source:
1. Rootman J. 2002. *Diseases of the orbit. A multidisciplinary approach.* 2nd ed. Philadelphia: Lippincott Williams and Wilkins.
2 Weill A , Cognard C, Castaings L, *et al.* Embolization of an orbital varix after surgical exposure. *Am J Neuroradiol* 1998;19:921–3.

65. Answer: c

Discussion: The rubella virus is a single stranded RNA virus. A recent systematic review concluded that the risk of intrauterine infection with maternal exposure was highest in both the first trimester (90%) and the last month of pregnancy (100%).
However, the risk of congenital defects is highest if infection occurs within the first trimester (100%), and virtually 0% in the 3rd trimester.
The commonest systemic manifestation is hearing impairment (44%) and the commonest ocular manifestation is salt and pepper retinopathy (22%) in which the ERG is normal.

Source:
1. Mets MB, Chhabra MS. Eye Manifestations of Intrauterine Infections and Their Impact on Childhood Blindness. *Surv Ophthal.* Vol 53(2) March – April 2008

66. Answer: b

Discussion: This is the classic 'swiss cheese' appearance of the cribiform variant of adenocystic carcinoma of the lacrimal gland.

Source:
1. Eagle, RC. 2011. *Eye Pathology: An Atlas and Text.* 2nd Ed. Lippincott Williams & Wilkins

67. Answer: b

Discussion: Congenital Brown's rarely resolves spontaneously, and surgery should be considered once a significant abnormal head posture appears or binocular single vision starts to deteriorate.
Acquired Brown's may occur as a result of trauma, orbital masses, or inflammation. As such, MRI and investigations to exclude masses, lupus and rheumatoid arthritis should be considered.

Source
1. Wright KW. *Color Atlas of Ophthalmic Surgery-Strabismus.* Philadelphia, Pa: Lippincott; 1991:201-219.
2. Wright KW, Silverstein D, Marrone AC, Smith RE. Acquired inflammatory superior oblique tendon sheath syndrome. A clinicopathologic study. *Arch Ophthalmol.* Nov 1982;100(11):1752-4

68. Answer: a

Discussion: Incontinentia Pigmenti is an X –linked dominantly inherited neurocutaneous disorder which is lethal in utero for males. It is characterised by red papules or vesicles at birth that regress over months to be replaced with reticulate or linear hyperpigmentation.

20-35% of children have ophthalmic manifestations, thought to arise from occlusions of the retinal vasculature, resulting in proliferative retinopathy, tractional retinal detachments and vitreous haemorrhage.

Source:
1. Holmstrom G, Thoren K. Ocular manifestations of incontinentia pigmenti. *Acta Ophthalmol Scand.* Jun 2000;78(3):348-53

69. Answer: b

Discussion: Mobius syndrome is a rare, sporadic, congenital syndrome characterised by hypoplastic 6th and 7th nerve nuclei causing bilateral 6th and 7th nerve palsies. In addition, there is often a restrictive component due to bilateral tight medial recti, causing a positive forced duction test,

Source:
1. Magli A, Bonavolonta F, Forte R, Vassalo P. Lower eyelid surgery for lagophthalmos in Möbius and Poland-Möbius syndromes. *J Craniofacial Surg.* 2011 Nov;22(6)53-4

70. Answer: a

Discussion: The normal, base in (abduction) fusional reserves at distance are 5-10 PD.

Source:
1. Stidwill D, Fletcher R. 2011. *Normal Binocular Vision: Theory, Investigation and Practical Aspects.* Oxford. Blackwell Publishing.

71. Answer: a

Discussion: Patients with solar retinopathy most often present within days post exposure, with central scotoma, negative afterimage and reduced visual acuity. The damage most often occurs in the dominant eye but may be bilateral. The fundus may show a small foveolar yellow gray spot and macular oedema that subsides over a matter of days to weeks, with loss of foveal depression or pigmentary changes. True macular holes are rarely seen.

Source:
1. Ferenc Kuhn. 2008. *Ocular Traumatology*. Springer.

72. Answer: c

Discussion: Surgical intervention is indicated in patients with full thickness macular holes. Prophylactic vitrectomy has not been shown to prevent full thickness macular hole formation in eyes with Stage 1 holes. Laser retinopexy has been associated with poor visual outcome due to adjacent thermal damage to the fovea.
Vitrectomy, internal limiting membrane peel and intraocular gas tamponade is the surgical method of choice. Eyes that stand to gain the most improvement are eyes with worse pre-operative visual acuity. The use of autologous platelets as a surgical adjunct has been shown to increase the rate of anatomic closure without any difference in the final post operative visual acuity.

Source:
1. de Bustros S. Vitrectom for prevention of macular holes: results of a randomised multicentre clinical trial. Vitrectomy for Prevention of Macular Hole Study Group. *Ophthalmology* 1994; 101:1055-1059
2. Cox MS. Discussion of Shocket SS, Lakhanpal V, Xiaoping M et al. Laser treatment of macular holes. *Ophthalmology* 1988:581-582
3.Paques M, Chastang C, Mathis A et al. Effect of autologous platelet concentrate in surgery for idiopathic macular hole: results of a mulcentre double masked, randomised trial. Platelets in macular hole study group. *Ophthalmology* 1999; 106:932-938

73. Answer: a

Discussion: A relative visual field defect is characteristic of a rhegmatogenous retinal detachment whereas absolute field defects are characteristic of retinoschisis.

Source:
1. Willamson TH. 2013. *Vitreoretinal Surgery*. 2nd Ed. Berlin Heidelberg. Springer-Verlag.

74. Answer: b

Discussion: The diagnosis of HZO is clinical. Investigations in the clinical setting are often not required, unless HZO is suspected in select patient groups (e.g HIV patients) in whom the presentation may be atypical.
The easiest test to perform is the Tzanck smear, which involves de-roofing vesicular lesions and spreading the fluid on a glass slide, allowing it to air dry, followed by Giemsa staining to look for multinucleated giant cells. However, the test is not sensitive, and a negative result should not preclude prompt antiviral therapy for patients in whom clinical suspicion is high.
DFA testing and polymerase chain reaction (PCR) of corneal scrapes or vesicular fluid carries a high sensitivity and specificity and are investigations of choice when a prompt, accurate diagnosis is required.

Source:
1. Ozcan A, Senol M, Saglam H, Seyhan M, Durmaz R, Aktas E, et al. Comparison of the Tzanck test and polymerase chain reaction in the diagnosis of cutaneous herpes simplex and varicella zoster virus infections. *Int J Dermatol.* Nov 2007;46(11):1177-9

75. Answer: d

Discussion: Phacoanaphylactic uveitis is a granulomatous inflammatory glaucoma historically occurring after trauma that disrupts the lens capsule, exposing the immunologically privileged lens protein. It is a predominantly Type 4 hypersensitivity reaction, histologically demonstrating a classic zonal granulomatous response with a zone of polymorphonuclear lymphocytes surrounded by a zone of macrophages and giant cells. Inflammation usually occurs after sensitization, and is most common within the first 2 weeks after trauma or cataract surgery with retained lens fragments.

Source:
1. Halbert SP, Manski W. Biological aspects of autoimmune reactions in the lens. *Invest Ophthal.* 1965;4:516-530.

76. Answer: a

Discussion: Aqueous misdirection typically occurs in patients with angle closure glaucoma who have undergone filtration surgery. Patients present with a red, aching eye, a flat anterior chamber and elevated intraocular pressure. Known precipitants have included post-operative use of miotics and cessation of cycloplegics. The anterior chamber typically remains flat despite a patent iridotomy.

77. Answer: b

Discussion: CHSD does not present with corneal guttata. Guttata in FHED and PPMD are likely to be bilateral (asymmetric in the case of PPMD) but unilateral in ICE.

78. Answer: b

Discussion: Conjunctival melanoma is an uncommon lesion. 70% arise from primary acquired melanosis 20% arise from a pre-existing nevus, and 10% arise de novo. Haematogenous and lymphatic spread are well documented while the lesion may also extend directly into the orbit.

79. Answer: c

Discussion: While silicone oil may cause angle closure glaucoma in a phakic eye – by virtue of pupil block, it more commonly causes a secondary open angle glaucoma by emulsification and blockage of the trabecular meshwork.

80. Answer: b

Discussion: By definition, PAM is an acquired unilateral, flat, freely moving, pigmented lesion often involving the bulbar conjunctiva, without a history of inciting trauma or inflammation. PAM may be further classified into *PAM with atypia* or *PAM without atypia*. PAM with atypia has been reported to have an associated risk of progression to melanoma of 36-75%. Nodularity is a sign of malignant change.

Source:
1. Zembowicz A, Mandal RV, Choopong P. (*2010*) Melanocytic Lesions of the Conjunctiva. *Archives of Pathology & Laboratory Medicine*: December 2010, Vol. 134, No. 12, pp. 1785-1792

81. Answer: a

Discussion: Gyrate atrophy is autosomal recessively inherited.

82. Answer: d

Discussion: The major cause of visual loss in patients with angioid streaks is secondary CNV formation, which occurs in 70-86% of patients. Other complications include foveal involvement by the streaks.

Source:
1. Piro PA, Scheraga D, Fine S. Angioid Streaks: Natural history and visual prognosis. In: *Management of Retinal Vascular and Macular Disorders*. Williams and Wilkins; 1983:136-9

83. Answer: a

Discussion: Meesman's Dystrophy is an innocuous corneal epithelial dystrophy characterized by multiple small central epithelial vesicles. Patients may develop photophobia and rarely, reduced vision in adulthood. The genes responsible are K3 and K12 that code for *keratin* proteins which form the epithelial cellular cytoskeleton.

Source:
1. Burns RP. Meesman's Corneal Dystrophy. *Trans Am Ophthalmol Soc*. 1968; 66: 530–635.

84. Answer: b

Discussion: The risk of developing radiation retinopathy (RR) is both dose and fraction size dependent. Parsons et al. determined that the risk of developing RR was in the region of 53% of patients receiving a total of 45-55 Gy. Independent studies have corroborated their findings and report that a minimum exposure of 30-35 Gy is required before RR can be expected. Total doses of 60 Gy and 70 Gy have been reported to cause retinal changes in 50% and 85-95% of patients respectively.

Source:
1. Archer DB. Doyne Lecture: responses of retinal and choroidal vessels to ionizing radiation. *Eye* 1993; 71-13
2. Nakissa N, Rubin P, Strohl R et al. Ocular and orbital complications following radiation therapy of paransal sius malignancies and review ofliterature. *Cancer* 1983; 51:980-986
3. Merriam GR Jr, Szechter A, Focht EF. The effects of ionizing radiations on the eye. *Front Radiat Ther Oncol* 1972; 6:346-385
4. Parsons JT, Bova FJ, Fitzgerald CR et al. Radiation retinopathy after external beam irradiation: analysis of time-dose factors. *Int J Radiat Oncol Biol Phys* 1994; 30:765-773

85. Answer: b

Discussion: The optic radiations pass from the lateral geniculate body (LGB) to synapse in the striate cortex where they may be injured either within the temporal, parietal lobes or the internal capsule. The radiations form the most posterior portions of the internal capsule. Signs of lesions affecting the radiations include a normal visual acuity and color vision (assuming unilateral lesions), homonymous hemianopic field defects that may be more dense inferiorly or superiorly (depending on the location of the lesion), and normal optic discs.
Damage to the internal capsule may also be associated with transient ocular deviation towards the side of the lesion.

Source:
1. Newman NJ, Miller NR, Biousse V. 2008. *Walsh and Hoyt's Clinical Neuro-Ophthalmology: The Essentials*. 2nd Ed. Lippincott Williams & Wilkins

86. Answer: a

Discussion: Features of keratoconus include progressive myopia, multiple inadequate spectacle corrections in part due to the progressive nature of the condition as well as the resulting irregular astigmatism. Videokeratography may reveal progressive thinning (involving the inferior cornea in 80% of patients) as well as increasing keratometric values.

Source:
1. Rabinowitz YS. Keratoconus. *Surv Ophthalmol*. Jan-Feb 1998;42(4):297-319.

87. Answer: a

Discussion: The findings above point to a diagnosis of a retinal capillary haemangioma, which may be associated with underlying Von Hippel Lindau disease. The lesions are typically situated between arteries and veins and may threaten vision by causing macular exudation or exudative retinal detachment. Fluorescein angiogram shows rapid filling of the feeding artery, followed by complete hyperfluorescence of the lesion, filling of the vein and diffuse late leakage.

In contrast, cavernous haemangiomas may show slow, incomplete filling with a fluid meniscus and no leakage while racemose haemangiomas show rapid filling with no leak.

88. Answer: c

Discussion: Bell's palsy is the commonest cause of unilateral LMN 7th palsy. It is strongly linked with herpes simplex. Spontaneous recovery is common and steroids increase the chances of recovery.

Source:
1. Anderson P. New American Academy of Neurology guideline on Bell's palsy. *Medscape Medical News*. November 7, 2012. Accessed November 12, 2012.
2. Gronseth GS, Paduga R. Evidence-based guideline update: Steroids and antivirals for Bell palsy: Report of the Guideline Development Subcommittee of the American Academy of Neurology. *Neurology*. Nov 7 2012
3. Murakami S, Mizobuchi M, Nakashiro Y, Doi T, Hato N, Yanagihara N. Bell palsy and herpes simplex virus: identification of viral DNA in endoneurial fluid and muscle. *Ann Intern Med*. Jan 1 1996;124(1 Pt 1):27-30.

89. Answer: b

Discussion: The paradoxical pupillary response, classically described by Barricks, is the initial constriction, followed by slow dilation of the pupils in response to dimming of the lights. It is seen in patients with rod monochromatism and congenital stationary night blindness (CSNB), and should alert the physician to a retinal dystrophy if present in a child with congenital nystagmus or nyctalopia.

Source:
1. Barricks ME, Flynn JT, Kushner BJ. Paradoxical pupillary responses in congenital stationary night blindness. *Arch Ophthalmol.* 1977 Oct;95(10):1800-4.
2. Brodsky MC. 2010. *Pediatric Neuro-Ophthalmology.* 2nd Edition. Springer

90. Answer: d

Discussion: Bias may arise from a failure of participants to complete a study, for example, due to adverse effects of any experimental therapy. This in turn leads to missing data, and may, for example, cause underestimation of frequency of adverse effects. Methods of minimizing bias caused by non-completion of the trial and subsequent missing data include Worse Case Scenario analysis, Intention to Treat analysis, and Last Observation Carried Forward.

Treatment allocation concealment is a method to minimize selection bias.

Source:
1. Wang D, Bakhai A. 2006. *Clinical Trials: A Practical Guide to Design, Analysis, and Reporting.* Remedica.

Paper 2

1. A 9-year-old boy with cleft palate, hearing aids and thick concave glasses presents with bilateral rhegmatogenous retinal detachments. Abnormalities in which gene are MOST likely to be implicated in this scenario?

a. COL2A1
b. COL11A2
c. COL9A2
d. VCAN

2. Of the following, which drug is LIKELIEST to be the culprit in the following scenario?

A 47-year-old gentleman with a facial wound and orbital cellulitis is treated with intravenous ceftriaxone and metronidazole. Cultures are positive for MRSA. He is then treated with intravenous clindamycin. De-colonization with mupirocin ointment is also initiated. Within 3 days he develops severe abdominal pain and diarrhoea.

a. clindamycin
b. ceftriaxone
c. metronidazole
d. mupirocin

3. Choose the TRUE statement regarding astigmatism.

a. Irregular astigmatism can be corrected satisfactorily with glasses
b. With accommodation relaxed, in simple myopic astigmatism, both focal lines are in front of the retina
c. Is typically against the rule in Pellucid Marginal Degeneration
d. With the rule astigmatism can be corrected with a minus cylinder with its axis at 180 degrees

4. The following may cause central retinal artery occlusions EXCEPT?

a. Protein C deficiency
b. Factor 5 Leiden
c. Antithrombin 3 deficiency
d. Liver cirrhosis

5. Which of the following genes is NOT implicated in the pathology of congenital stationary night blindness?

a. connexin 50
b. rhodopsin
c. rod cGMP-PDE
d. rod transducin

6. Which of the following parasite–vector pairs with ocular significance is FALSE?

a. *Onchocerca volvulus* – Blackfly
b. *Loa – loa* – Deerfly
c. *Schistosoma* – Fish
d. *Leishmania* - Sandfly

7. The following mucopolysaccharidoses (MPS) are autosomal recessively inherited EXCEPT?

a. Hurler-Scheie
b. Sanfilipo
c. Hunter
d. Sly

8. Which statement regarding posterior lenticonus is MOST likely to be true?

a. Not associated with Alport Syndrome
b. Presents with cataract in adulthood
c. Usually unilateral
d. Girls are more likely to be affected

9. What is the LIKELIEST diagnosis for the following scenario?

A 43-year-old lady complains of blurring of vision and haloes in her right eye. On examination, her vision is OD: 6/48 OS: 6/9 uncorrected. The intraocular pressure (IOP) is OD: 40 OS: 16 There are corneal guttata, an irregularly shaped pupil and high peripheral anterior synechiae in her right eye. Her left eye is normal.

a. Fuchs Heterochromic Iridocyclitis (FHIC)
b. Herpetic Uveitic Glaucoma
c. Iridocorneal – Endothelial Syndrome (ICE)
d. Schwartz Syndrome

10. Which of the following statements about the natural history of branch retinal vein occlusion (BRVO) is TRUE?

a. The prognosis in terms of visual acuity depends on the presenting visual acuity
b. The majority of patients present with a visual acuity of 6/36 or worse
c. The majority of patients experience significant visual improvement
d. A minority of patients develop macular oedema over the course of a year

11. Regarding visual field defects in non-arteritic anterior ischaemic optic neuropathy, which defect is MOST likely to occur?

a. Altitudinal
b. Central
c. Hemianopic
d. Enlarged blind spot

12. Which of the following cutaneous lesions is LEAST likely to be associated with sun exposure?

a. Seborrhoeic keratosis
b. Squamous Cell Papilloma
c. Keratoacanthoma
d. Basal Cell Carcinoma

13. Regarding the following scenario, which is the MOST likely diagnosis?

A 3-year-old boy presents with an alternating esodeviation. According to his mother, it has been present since birth.
On examination, he has an abnormal head posture with a left face turn. When his head is straightened, he has horizontal nystagmus in the primary gaze position.
On motility testing, he demonstrates cross fixation with the esodeviation most prominent in lateral gaze to either side. The nystagmus appears on abduction of either eye.

a. Congenital Nystagmus
b. Vestibular nystagmus
c. Nystagmus Block Syndrome
d. Congenital Esotropia

14. The following White Dot Syndromes have a female predilection EXCEPT?

a. Multiple Evanescent White Dot Syndrome (MEWDS)
b. Punctate Inner Choroidopathy (PIC)
c. Serpiginous Choroidopathy
d. Birdshot Choroidopathy

15. Which of the following hereditary vitreoretinopathies is NOT associated with an increased risk of retinal detachment?

a. Jansen's Disease
b. Wagner's Disease
c. Familial Exudative Vitreoretinopathy (FEVR)
d. Stickler's Syndrome

16. A 21-year-old rugby player presents with bilateral mucopurulent discharge, follicular conjunctivitis and lymphadenopathy. There are multiple subepithelial infiltrates. He gives a history of multiple casual sexual encounters. The following are relevant investigations for the abovementioned scenario EXCEPT?

a. Conjunctival Scrapes for McCoy culture and Giemsa stain
b. Conjunctival Scrapes for Direct Immunofluorescence
c. Conjunctival Scrapes for Enzyme Linked Immunofluorescence (ELISA)
d. Conjunctival Scrapes for Gram Stain

17. Which is likely to be the MOST helpful investigation for the following scenario?

A 30-year-old gentleman who has a history of being assaulted with multiple punches to his face complains of a watering right eye.
Probing and syringing reveals a hard stop with reflux from both puncta and no saline entering the nose.

a. Computed tomography (CT) scan of the midface
b. Lacrimal Scintigraphy
c. Dacryocystorhinogram (DCG)
d. Combined CT of the midface and DCG

18. Concerning computed tomographic (CT) scans, which of the following statements is MOST likely to be false?

a. Non-contrasted CT scans are excellent for imaging calcification and bone
b. Requires contrast to demonstrate an orbital haemorrhage
c. Utilises ionizing radiation
d. Details of the optic nerve are better with an MRI

19. Which of the following is likely to be the BEST option to determine the intraocular pressure in a patient with heavily scarred corneas?

a. Pneumotonometry
b. Tono-pen
c. Rebound Tonometry
d. Applanation Tonometry

20. Which of the following statements about the Ishihara Pseudoisochromatic plates is MOST likely to be true?

a. Relies on difference in hue to distinguish the different numbers from surrounding dots
b. Able to detect tritanopia
c. Should be read indoors under fluorescent lighting
d. Prolonged exposure of the test to sunlight may cause loss of accuracy of test readings

21. Which of the following statements is MOST likely to be false regarding imaging studies in thyroid eye disease?

a. Magnetic Resonance Imaging (MRI) is better than Computed Tomography (CT) in detection of optic nerve compression
b. Isolated single rectus muscle involvement is a rarity
c. Tenting of the posterior aspect of the globe is associated with a poorer visual outcome
d. Orbital fat density on CT scans is often hypodense to that of normal patients

22. According to the results of the following investigations, what is the MOST likely diagnosis?

A flat, orange – yellow, well demarcated choroidal lesion in both eyes. It has pseudopod – like edges. Ultrasound shows a highly reflective anterior border with orbital shadowing. CT scan shows bilateral dense lesions at the level of the choroid. MRI reveals hyperintense T1 signals and hypointense T2 signals.

a. Choroidal Metastasis
b. Choroidal Osteoma
c. Choroidal Melanoma
d. Choroidal Haemangioma

23. Regarding tests of binocular single vision, which of the following statements is MOST likely to be false?

a. A Worth Four Dot test where the patient reports seeing three lights reveals suppression of the eye behind the red glass
b. Bagolini Striated glasses are placed with their axes at 90 and 180 degrees
c. The 4-dioptre prism test is primarily used to diagnose monofixation syndrome
d. The afterimage test is used to determine if a patient has normal or anomalous retinal correspondence

24. The term *pharmacodynamics* refers to

a. The application of a drug to achieve a given endpoint
b. The biological activity of the drug, including receptor binding and subsequent cellular effects
c. The drug cycle through the body, including metabolism and excretion
d. The side effects of the drug

25. Choose the INCORRECT statement regarding phenylephrine.

a. Phenylephrine is a direct acting alpha-1 agonist
b. Phenylephrine mydriasis alone is usually sufficient for dilated fundoscopy
c. The use of phenylephrine 2.5% may be hazardous in infants
d. Phenylephrine should be used with caution in patients on antidepressants

26. A 10-year-old girl with keratoconus presents to your clinic. She has a visual acuity of 6/6 in both eyes and progressive astigmatism in her right eye that is still correctable with rigid gas permeable contact lenses.
What would be the treatment option of choice?

a. Rigid Gas Permeable (RGP) contact lenses
b. Spectacles
c. Collagen Cross Linking (CCL)
d. Anterior Lamellar Keratoplasty

27. Which of the following statements about antimetabolite use in glaucoma surgery is MOST likely to be false?

a. Mitomycin C (MMC) should be applied beneath the scleral flap prior to entering the anterior chamber
b. As per NICE Glaucoma Guidelines, is indicated in all patients with primary open angle glaucoma undergoing trabeculectomy
c. Should be avoided in pregnancy
d. Post operative needling is typically done with a total dose of 5-Fluorouracil (5-FU) 10mg

28. Which hypersensitivity reaction is MOST likely to be implicated in phacoanaphylactic glaucoma?

a. Type 1
b. Type 2
c. Type 3
d. Type 4

29. Which of the following statements regarding the pathophysiology and presentation of ocular cicatricial pemphigoid (OCP) is MOST likely to be correct?

a. Cutaneous manifestations are the norm
b. OCP is characterized by a Type 2 hypersensitivity reaction towards basement membrane proteins
c. There is no sex predilection
d. The majority of patients are in their 30s

30. Which of the following statements regarding Peripheral Ulcerative Keratitis (PUK) is MOST likely to be true?

a. PUK occurs more commonly in men
b. It happens commonly in patients with Relapsing Polychondritis (RP)
c: The commonest systemic disorder associated with PUK is rheumatoid arthritis
d. It is a predominantly Type 2 hypersensitivity reaction

31. Regarding pigment dispersion syndrome (PDS), which of the following statements is MOST likely to be true?

a. PDS affects men more than women
b. PDS associated glaucoma occurs earlier in men than in women
c. The syndrome is associated with peri pupillary iris atrophy
d. Is often unilateral

32. Which of the following is LEAST likely to be associated with sub-retinal haemorrhage?

a. Presumed Ocular Histoplasmosis (POHS)
b. Choroidal Rupture
c. Diabetic Retinopathy
d. Best's Disease

33. Which of the following statements regarding achromatopsia is INCORRECT?

a. Is inherited in an autosomal recessive manner
b. Nystagmus is often congenital or develops few weeks after birth
c. Is most commonly incomplete
d. The majority of patients exhibit a normal fundus

34. The following drugs may cause cystoid macular oedema EXCEPT?

a. Topical latanoprost
b. Tamoxifen
c. Intravenous interferon
d. Desferrioxamine

35. According to Royal College guidelines for Intravitreal Injections, which of the following statements is TRUE?

a. Injections can be given 3.0 mm from the limbus in phakic patients
b. IOP check is mandatory in all patients
c. Can be given in an outpatient clinic setting dealing with casualty cases
d. If the patient has no perception to light (NPL) post injection, an anterior chamber paracentesis should be considered

36. Regarding obtaining consent for a patient undergoing cataract surgery, which of the following statements is FALSE?

a. A spouse is not able to give consent on behalf of an adult patient who is incapacitated from advanced Alzheimer's
b. When obtaining consent, postoperative endophthalmitis must be mentioned as a potential complication
c. Parental consent must be obtained for a 1-year-old child with congenital cataract
d. It is the duty of the doctor to estimate and weigh the amount of information the patient is likely to want to hear to decide how much information to give them

37. Regarding sensitivity and specificity, choose the CORRECT statement.

a. The sensitivity of a test is directly proportional to its false negative rate
b. The sensitivity of a test is calculated as the proportion of patients testing positive divided by the number of true positives and true negatives.
c. The specificity of a test is calculated as the proportion of patients testing negative divided by the number of patients without the disease
d. A lower specificity is desirable for a confirmatory test

38. Which of the following statements regarding Catscratch Disease (CSD) is FALSE?

a. Is caused by a gram positive bacillus
b. May cause a unilateral conjunctivitis
c. Causes regional lymphadenopathy within 2 weeks from inoculation
d. Resolves spontaneously within 2-4 months

39. Regarding treatment options of non-arteritic anterior ischemic optic neuropathy (NA-AION) which of the following statements is LEAST likely to be true?

a. There is data to support steroid use once giant cell arteritis has been excluded
b. Optic nerve fenestration is not an option
c. Second episodes of NA-AION do not occur in an already involved eye
d. Risk of contralateral eye involvement is 19% within 5 years

40. Regarding the presenting features of Leber's Hereditary Optic Neuropathy (LHON), which of the following statements is MOST likely to be false?

a. Presentation is typically in otherwise healthy young adult males between the ages of 15-35
b. Sudden, sequential, painless blurring of vision is typical
c. There is often a family history where the patient's father has had similar symptoms
d. Point mutations coding for respiratory chain enzymes are responsible

41. Regarding Neuromyelitis Optica, which of the following statements is LEAST likely to be true?

a. Is an idiopathic, demyelinating disorder frequently involving the optic nerves and spinal cord
b. Is characterized by perivascular infiltration of neutrophils
c. Diagnosis is aided by the presence of antibodies to membrane phospholipids
d. Is characterized by a more frequently relapsing course as compared to classical multiple sclerosis

42. Which of the following statements regarding cutaneous squamous cell carcinoma (SCC) is MOST likely to be false?

a. Is the most common head and neck cutaneous malignancy
b. A tumor diameter of more than 2 cm is associated with an increased risk of disease related death
c. Patients presenting with associated regional paraesthesia and tingling, or muscular weakness are at increased risk of disease related death
d. Eyelid SCC is more common on the lower lid

43. Which of the following conditions is MOST likely to be the correct diagnosis for the following scenario?
A newborn baby presents with a unilateral droopy upper eyelid. Her mother says the drooping has been present since birth. The drooping seems to disappear when she starts to feed.

a. Marcus Gunn Jaw Wink
b. Isolated Congenital Ptosis
c. Floppy Eyelid Syndrome
d. Blepharophimosis Syndrome

44. Which of the following statements regarding Cyclical Esotropia is CORRECT?

a. Onset in adulthood
b. Demonstrates a manifest deviation followed by no deviation in cycles of constant duration
c. Angle at distance often significantly larger than angle at near
d. Amblyopia is common

45. The following are features of retained copper foreign body EXCEPT?

a. May present immediately post-trauma with an acute, fulminant endophthalmitis – like picture
b. Glaucoma
c. Greenish iris discolouration
d. Results in an electronegative electroretinogram (ERG)

46. Regarding white without pressure, which statement is MOST likely to be true?

a. Presents as a whitish sheen of the retina on scleral indentation
b. Choroidal markings are usually visible
c. Seen predominantly in young Caucasian patients
d. May be confused with shallow retinal detachments

47. Select the MOST appropriate treatment option for the following scenario.

A 65-year-old lady with hypertension and Type 2 diabetes mellitus presents with sudden onset blurring of vision in her right eye for 1 week. On examination, her best corrected visual acuity is OD 6/18 OS 6/9 with no RAPD. She has early cataract in both eyes. Intraocular pressure is 12 in both eyes.
The fundus of the left eye is normal with a cup disc ratio of 0.3.
The fundus of the right eye reveals haemorrhages along the superotemporal vein which is dilated. The macula is swollen with no foveal bleed. The cup disc ratio is 0.3.
OCT shows central foveal thickness of 400μm.

a. Intravitreal dexamethasone implant
b. Intravitreal ranibizumab monthly
c. Intravitreal triamcinolone
d. Macular grid laser

48. Regarding prismatic correction of adult onset diplopia, which of the following statements is MOST likely to be true?

a. A ground in prism is the option of choice for temporary management of diplopia caused by microvascular sixth nerve palsies
b. Fresnel prisms should be applied to both eyes
c. Ground in prisms of 12 dioptres or more are inferior to Fresnel prisms of similar power
d. The non-dominant eye is often fitted with Fresnel prisms

49. Pertaining to Valganciclovir, which is the statement MOST likely to be true?

a. Is less efficacious than ganciclovir
b. Causes less neutropenia than ganciclovir
c. Is administered intravenously
d. Has excellent bioavailability

50. According to the following scenario, which is the surgical procedure of choice?

A 49-year-old African immigrant presents with heavily vascularised corneas and poor vision. The lid margins of his upper eyelids are internally rotated. There is no lid retraction. Eversion of the upper lids reveals linear subconjunctival fibrotic bands and shortening of the posterior lamellae. There is no active inflammation.

a. Everting Sutures
b. Cryotherapy of the Eyelashes
c. Anterior Lamellar Repositioning
d. Tarsal Fracture and posterior lamellar graft

51. According to NICE guidelines in respect to ocular hypertension, which statement is FALSE?

a. All patients with a central corneal thickness (CCT) of less than 555μm should be treated
b. Patients with an intraocular pressure (IOP) of more than 32 mmHg should be treated regardless of the CCT
c. Patients with a CCT of >590μm and an IOP of >32 mmHg should be observed
d. Patients with a CCT of 555-590 mmHg and an IOP of <25 mmHg should be treated

52. Regarding the following scenario regarding patient confidentiality, which is the BEST course of action?

Mr. A, a 25-year-old gentleman, is seen in your Uveitis clinic with what appears to be cytomegalovirus retinitis. He is offered HIV testing, and the test comes back positive. He is married with 2 young children. He is very hesitant about breaking the news to both his wife and his GP.

a. Accept Mr. A's hesitation as perfectly normal, and respect his confidentiality by not saying anything to anyone
b. Encourage him to break the news to his wife, citing her need to be tested as her safety is at stake
c. Overlook his hesitation and arrange for your secretary to make an appointment with both his wife and GP without his permission
d. Ask him for permission for you to break the news to his wife

53. The following table contains data from a cross sectional study studying the association between cataract surgery with underlying diabetes and the development of post operative cystoid macular oedema (CMO).

	Development of CMO	No CMO	Total
Diabetes Present	30	10	40
Diabetes Absent	25	25	50

What is the odds ratio (OR) for development of CMO in patients with diabetes compared to patients without diabetes?

a. 0.3
b. 3
c. 0.03
d. 1.3

54. Which of the following cataract types is typically seen in atopic dermatitis?

a. Posterior Subcapsular Cataract
b. Anterior Subcapsular Cataract
c. Cortical Cataract
d. Nuclear Sclerotic Cataract

55. A 69-year-old lady undergoes routine cataract surgery that becomes complicated with a ruptured posterior capsule. A posterior chamber lens with a power of +23.00 dioptres was to be implanted. As the capsular rupture is extensive, the decision to implant a lens in the sulcus is made.

What should the new lens power be in order to attain the targeted refractive state?

a. +22.00
b +22.50
c. +23.50
d. +24.00

56. Which of the following statements regarding viscoelastics is MOST likely to be false?

a. Less viscous materials coat surfaces more effectively
b. Endothelial protection is more marked with more viscous viscoelastics
c. The higher the viscosity, the easier it is removed during aspiration
d. Air trapping is more common with less viscous viscoelastics

57. Regarding tilted discs, which of the following statements is MOST likely to be true?

a. Often bilateral, but asymmetrical
b. Associated with hypermetropia
c. May cause progressive visual field defects
d. The visual field defects are often indistinguishable from those caused by chiasmal lesions

58. Which of the following statements regarding filamentary keratitis is MOST likely to be false?

a. Is most often seen as a complication of keratoconjunctivitis sicca
b. Filaments consist of mucus threads attached to areas of punctate erosions
c. Normal saline drops may aid resolution of the disease
d. N-acetylcysteine is often helpful

59. Which of the following statements regarding orbital and adnexal lymphoma (OAL) is MOST likely to be false?

a. Are predominantly low-grade B cell lymphomas
b. Associated with primary central nervous system (CNS) lymphoma
c. Not known to be associated with Human Immunodeficiency Virus (HIV)
d. Radiotherapy, and not chemotherapy, is the mainstay of therapy

60. According to the following scenario, what is the MOST likely location of the offending lesion?

A 49-year-old gentleman presents with bilateral painless, progressive, blurring of vision. His visual acuity is OD: 6/24 OS 6/36. There is no relative afferent pupillary defect. On examination of the fundi, the temporal and nasal portions of the optic discs are pale.

a. Optic Nerve
b. Optic Chiasm
c. Optic Tracts
d. Internal Capsule

61. Which of the following disorders is NOT a cause of intermittent proptosis?

a. Orbital Lymphangioma
b. Capillary Haemangioma
c. Cavernous Haemangioma
d. Orbital Mucocele

62. Regarding congenital cytomegalovirus (CMV) infection, which of the following statements is MOST likely to be true?

a. Less common than congenital infection with rubella
b. Infection usually occurs at time of delivery
c. Chorioretinitis occurs in 20% of infected neonates
d. A course of intravenous ganciclovir is effective in eliminating all trace of the virus

63. What is the risk of developing systemic lymphoma within 5 years in a patient with orbital atypical lymphoid hyperplasia?

a. 10%
b. 20%
c. 30%
d. 40%

64. According to Huber's Classification of Duane's Syndrome, what proportion of all cases are Type 1?

a. 55%
b. 65%
c. 75%
d. 85%

65. The following ocular features are typical findings in Fetal Alcohol Syndrome (FAS) EXCEPT?

a. Telecanthus
b. Short horizontal palpebral fissures
c. Myopia
d. Glaucoma

66. Which of the following is NOT an indication of botulinum toxin use?

a. Correction of diplopia in 6th nerve palsies
b. Correction of hypotropia in patients with quiescent thyroid eye disease
c. Congenital Esotropia
d. Intermittent Exotropia

67. What are the typical electroretinogram (ERG) changes in progressive ocular siderosis?

a. Early loss of both the a and b waves
b. An early supranormal a wave, followed by persistent, progressive loss of the b wave
c. Early loss of the b wave
d. Normal ERG

68. Which of the following statements regarding juvenile retinoschisis is TRUE?

a. Affects boys and girls equally
b. Usually affects peripheral retina
c. Is not associated with systemic findings
d. Prophylactic treatment to prevent retinal detachment is recommended

69. Which of the following is NOT a risk factor for band keratopathy?

a. Uncontrolled glaucoma
b. Hypoparathyroidism
c. Vitamin D toxicity
d. Ichthyosis

70. Which of the following statements regarding phacolytic glaucoma is MOST likely to be false?

a. Typically occurs in a hypermature cataract
b. The lens capsule is typically not intact
c. Occurs more commonly in underdeveloped countries
d. There is no gender bias

71. The following statements regarding medical options for treatment of aqueous misdirection syndrome are true, EXCEPT?

a. Topical atropine increases zonular tension, flattening of the lens and posterior movement of the lens iris diaphragm
b. Topical pilocarpine increases aqueous flow through the trabecular meshwork, decreasing intraocular pressure
c. Hyperosmolar agents decrease vitreous volume
d. Beta blockers act by reducing aqueous production

72. Regarding Lattice Dystrophy, choose the INCORRECT statement.

a. Is associated with mutations of the BIGH9 gene
b. Types 1 and 2 present with recurrent corneal erosions, whilst Type 3 tends to present with painless blurring of vision
c. Cardiac failure is an associated systemic feature
d. Recurrences are common after keratoplasty

73. Which is the LIKELIEST diagnosis for the following scenario involving a pigmented conjunctival lesion?

A 17-year-old boy of Afro Caribbean background presents with bilateral, patchy, freely moving conjunctival pigmentation dispersed along the limbus in both eyes. It has been there since birth, with no increase in size.

a. Conjunctival Melanoma
b. Congenital Ocular Melanosis
c. Conjunctival Melanocytoma
d. Benign Conjunctival Melanosis

74. When does choroideremia usually present in females?

a. 1st decade of life
b. 2nd decade of life
c. 3rd decade of life
d. Asymptomatic

75. As per Royal College Guidelines on Blood Borne viral infections, which of the following statements is MOST likely to be true?

a. All surgeons performing cataract surgery should be tested for HIV
b. Surgeons performing orbital surgery must be tested for HIV, Hepatitis B and C
c. All surgeons who are Hepatitis B positive should be barred from performing cataract surgery
d. All surgeons who are Hepatitis B positive should be barred from performing orbital surgery

76. Which of the following disorders is LEAST likely to be associated with angioid streaks?

a. Rheumatoid Spondylitis
b. Paget's Disease
c. Sickle Cell Disease
d. Pseudoxanthoma Elasticum

77. Which of the following is NOT a sign of quiescent interstitial keratitis (IK)?

a. Ghost vessels
b. Corneal thinning
c. Stromal salmon patch
d. Corneal scarring

78. The following increase the risk of developing radiation retinopathy in a patient undergoing cephalic radiotherapy for nasopharyngeal carcinoma EXCEPT?

a. Diabetes
b. Concurrent Chemotherapy
c. Higher total dose
d. Greater number of fractions for a given total dose

79. According to the following scenario, what is the LIKELIEST site of the offending lesion?

A 60-year-old gentleman presents with sudden onset binocular diplopia. His visual acuity is 6/6 in both eyes.
There is a left esotropia in primary gaze. He is unable to abduct his left eye past the midline.
There is no ptosis, and pupil reactions are normal. Optic discs are normal. There are no other cranial neuropathies. He also has weakness of the right side of his body.

a. Dorsal pons
b. Ventral pons
c. Dorsomedial pons
d. Dorsolateral pons

80. Which of the following is NOT a feature of Terrien's Marginal Degeneration?

a. Painless
b. Bilateral, progressive with the rule astigmatism
c. Lipid line posterior to area of thinning
d. Preferentially affects men

81. The following statements regarding the mode of spread by choroidal melanoma are true EXCEPT?

a. Haematogenous metastasis is most commonly to the liver
b. Optic nerve extension is common
c. Orbital invasion occurs via a transcleral route
d. Lymphatic spread is not possible

82. Regarding intracranial aneurysms, which of the following statements is MOST likely to be false?

a. Mortality in associated sub - arachnoid haemorrhage (SAH) is reported to be as high as 65%
b. Anterior communicating artery aneurysms may cause fluctuating central visual loss
c. Intracavernous aneurysms typically present with a temporal field defect
d. Middle cerebral artery (MCA) aneurysms may present with a homonymous hemianopia

83. Which of the following is MOST likely to present with progressive nyctalopia in childhood?

a. Fundus Albipunctatus
b. Stargardt's Disease
c. Refsum's Disease
d. Fundus Flavimaculatus

**84. A 60-year-old lady with thyrotoxicosis and poorly controlled schizophrenia presents with sudden onset proptosis of her right eye with exposure keratopathy and corneal breakdown.
Urgent CT scans reveal muscular enlargement with tendon sparing.**

What is the MOST appropriate treatment option?

a. Pulsed Intravenous Methylprednisolone
b. Oral Prednisolone
c. Surgical Decompression
d. Radiotherapy

85. Pertaining to the prognosis of HIV infection, which statement is MOST likely to be correct?

a. Mortality in non-treated individuals is 50%
b. Mortality tends to be higher in patients infected via intravenous needle sharing as compared to patients infected via other routes
c. HAART reduces the incidence of HIV/AIDS associated lymphoma
d. Non-nucleoside Reverse Transcriptase inhibitors (NNRTI) are effective against HIV-2

86. Which of the following conditions is NOT associated with difficulty in initiating lid opening?

a. Parkinson's Disease
b. Progressive Supranuclear Palsy (PSP)
c. Huntington's Disease
d. Gullian – Barre Syndrome

87. The following drugs are likely to cause nystagmus EXCEPT?

a. Gabapentin
b. Phenytoin
c. Barbiturates
d. Sodium valproate

88. Which is likely to be the investigation of choice in the following scenario?

A 15-year-old boy was in a car accident where he suffered a blow to his chest. He presented a week after the accident with chest pain radiating to his back. On examination, his left pupil is smaller than his right. The anisocoria is worse in the dark. Immediate cocaine testing is positive.

a. Magnetic Resonance Imaging (MRI) of the brainstem
b. Echocardiogram
c. Contrasted CT scan of the brainstem
d. CT Angiographic scan of the thoracic vessels

89. Which of the following statements regarding Ankylosing Spondylitis (AS) is MOST likely to be false?

a. 90% of patients with AS are HLA B27 positive
b. AS tends to affect young men
c. When AS affects women, the disease is frequently more crippling
d. Acute anterior uveitis occurs in 20-30% of patients with AS

90. The failure to prioritize publication of well-designed trials with negative or equivocal outcomes is called

a. Selection Bias
b. Publication Bias
c. Ascertainment Bias
d. Pre-Trial Bias

Paper 2 Answers and Discussion

1. Answer: a

Discussion: The above scenario is likely to point to a diagnosis of Stickler syndrome. Mutations in genes for collagen 2, 11 and 9 have been implicated. However, mutation of COL2A1 is responsible for 80-90% of cases. Mutations in VCAN are responsible for Wagner's, which bears some similarities to Stickler but lacks systemic features.

Source:
1. Kloeckener-Gruissem B, Amstutz C, VCAN Related Vitreoretinopathy. *Gene Reviews*. Pagon RA, Adam MP, Bird TD, et al., editors. Seattle (WA): University of Washington, Seattle 1993-2013.2.
2. Robin N.H, Moran R.T, Warman M, Ala-Kokko L. Stickler Syndrome. *Gene Reviews*. Pagon RA, Adam MP, Bird TD, et al., editors. Seattle (WA): University of Washington, Seattle 1993-2013.

2. Answer: a

Discussion: While pseudomembranous colitis may occur with use of any broad-spectrum antibiotic, it is classically linked to clindamycin.

3. Answer: c

Discussion: Irregular astigmatism is often difficult to correct satisfactorily with glasses due to meridians of greatest and least power not being perpendicular to each other. Often, rigid gas permeable contact lenses or refractive surgery (but not in the case of ectasias!) suffice.
In simple myopic astigmatism, one focal line falls in front of the retina, and one falls on the retina.
Astigmatism is often against the rule in Pellucid and Terrien degeneration.
With the rule astigmatism can be corrected with a minus cylinder with its axis at 180 degrees

Source:
1. Wilson SE, Lin DTC, Klyce SD, Insler MS: The corneal topography of Terrien's marginal degeneration. *Refract Corneal Surg* 6:15–20, 1990
2. Maguire LJ, Klyce SD, McDonald ME, Kaufmann HE: Corneal topography of pellucid marginal degeneration. *Ophthalmology* 94:519–524, 1987

4. Answer: d

Discussion: Thrombophilia may be caused by a deficiency of any of the regulatory factors required to avoid spontaneous clot formation.
Protein C and protein S degrade activated factors 5 and 7, and therefore, a deficiency in either may lead to a thrombophilic state.
Factor 5 Leiden is a mutant factor resistant to inactivation by activated Protein C and S.
Liver cirrhosis impairs the ability of the body to produce mature clotting factors, partially by causing impairment of vitamin K metabolism.

5. Answer: a

Discussion: Mutations in the gene coding for connexin50 protein are implicated in inherited cataract, whilst rhodopsin, rod transducin, and cGMP-PDE are all involved in the phototransduction cycle.

Source:
1. Berry V. Connexin 50 mutation in a family with congenital "zonular nuclear" pulverulent cataract of Pakistani origin. *Hum Genet*. 1999 Jul-Aug;105(1-2):168-70.

6. Answer: c

Discussion : The range of ophthalmologically significant parasites is wide. Often thought of as 'exotic' disease or infection, they are nevertheless significant causes of blindness and subsequent socioeconomic burden. Both parasites and helminths may cause ocular infection.
Ophthalmologically significant protozoa include *Toxoplasma, Acanthamoeba, Giardia lamblia, Trypanasoma cruzi, Leishmania donovani,* and *Plasmodium falciparum.*
Ophthalmologically significant helminths include *Onchocerca volvulus, Toxocara canis, Schistosoma*(carried by water snails)*, Taenia solium* and *Loa-Loa*

Source:
1. Klotz SA, Penn CC, Negvesky GJ, Butrus SI. Fungal and Parasitic Infections of the Eye. *Clin Microbiol Rev.* 2000 October; 13(4): 662–685.

7. Answer: c

Discussion: All the MPS are recessively inherited except MPS 2 (Hunter), which is X Linked

Source:
1. Denniston AKO, Murray PI. 2006. *Oxford Handbook of Clinical Ophthalmology.* Oxford. Oxford University Press.

8. Answer: c

Discussion: Posterior lenticonus refers to conical bulging of the posterior lens surface. It may be inherited (x linked or dominant) or sporadic. It is associated with Alport's syndrome, Duane syndrome and Lowe's syndrome. They occur typically in males, are unilateral, and present with progressive cataract in childhood. They are associated with very thin capsules, and capsular polishing is not advised during surgery due to increased risk of capsular rupture.

Source:
1. Vedantham V Rajagopal J,. Ratnagiri PK Bilateral simultaneous anterior and posterior lenticonus in Alport syndrome. *Indian J Ophthalmol.* 2005 Sep;53(3):212-3.
2. Cheng KP, Hiles DA, Biglan AW, Pettapiece MC. Management of Posterior Lenticonus. . *J Pediatr Ophthalmol Strabismus.* 1991 May-Jun;28(3):143-9

9. Answer: c

Discussion: While both Herpetic Uveitis and ICE may present with a unilateral raised IOP, synechiae in association with an irregular pupil and guttata is not typical of herpetic uveitic glaucoma.
FHIC does not present with synechiae. Schwartz syndrome is associated with rhegmatogenous retinal detachment.

10. Answer: d

Discussion: The prognosis in terms of visual acuity in BRVO does not depend on the presenting vision.
A systematic review of 24 studies on BRVO revealed that the presenting vision tended to be 6/12 or worse.
Up to 15% of patients developed macular oedema over the course of a year.
Up to 50% of untreated patients fail to show any form of significant visual improvement.
20% show deterioriation in their vision.

Source:
1. Royal College Interim Guidelines for Retinal Vein Occlusion 2010

11. Answer: a

Discussion: Altitudinal field defects are the most common defects seen in non-arteritic anterior ischaemic optic neuropathy.

Source:
1. Hayreh SS. Posterior ischaemic optic neuropathy: clinical features, pathogenesis, and management. *Eye* (2004) 18, 1188–1206

12. Answer: b

Discussion: Generally, cutaneous lesions affecting the head, neck and eyelids associated with sun exposure can be divided into benign, premalignant and malignant lesions. Benign lesions include seborrhoeic keratosis. Premalignant lesions include actinic keratosis and keratoacanthomas. Malignant lesions associated with sun exposure include basal cell and squamous cell carcinomas as well as melanomas.
Squamous cell papillomas do not have any known cause and are the commonest benign lesions of the head and neck.

13. Answer: c

Discussion: Nystagmus Block Syndrome (NBS) develops in patients with congenital nystagmus. It is characterized by an esodeviation which aims to enhance visual acuity by dampening the nystagmus and stabilizing the foveal image. As the fixating eye adducts to dampen the nystagmus, the face turns in the opposite direction to allow fixation.
Motility is full, but children are often reluctant to abduct their eyes as it results in worsening of the nystagmus.

Source:
1. MacEwen C, Gregson R. 2003. *Manual of Strabismus Surgery*. London. Elsevier Limited.

14. Answer: c

Discussion: The White Dot Syndromes affect women more than men except in the case of Serpiginous Choroidopathy and Acute Posterior Multifocal Placoid Pigment Epitheliopathy (APMPPE) which affect men and women with equal frequency.

15. Answer: b

Discussion: An approach to rhegmatogenous retinal detachments should always take into consideration a hereditary cause. Of the hereditary vitreoretinopathies, Stickler's, Jansen and FEVR are associated with increased risk of rhegmatogenous retinal detachment, and in the case of FEVR, exudative and tractional retinal detachment.

16. Answer: d

Discussion: Adult inclusion conjunctivitis has been traditionally diagnosed by cultures and Giemsa stain to look for basophilic intracellular inclusion bodies. More recently, Direct Immunofluorescence and ELISA kits have become common-place. The bacteria are classified as gram negative, but stain poorly with Gram.

17. Answer: d

Discussion: A CT scan in association with a DCG would be able to accurately diagnose the presence of any midface fractures as a cause of duct obstruction.

18. Answer: b

Discussion: The demonstration of extravasation of blood does not require contrast.

19. Answer: a

Discussion: Accuracy of applanation tonometry is heavily influenced by corneal factors, notably, rigidity and thickness. Rebound tonometry is also influenced by corneal factors.
Measuring intraocular pressure in patients with heavily scarred or irregular corneas may be better done with pneumotonometry or the tonopen which are both less influenced by corneal factors. However, the results obtained with the tonopen are less reproducible as compared to pneumotonometry.

Source:
1. Giaconi J.A, Law S.K, Caprioli J, Coleman A.L. 2010. *Pearls of Glaucoma Management*. Springer
2. Marini M, Da Pozzo S, Accardo A, Canziani T. Comparing applanation tonometry and rebound tonometry in glaucomatous and ocular hypertensive eyes. *Eur J Ophthalmol* 2010 Sep 29;21(3):258-263
3. Molina N, Milla E, Bitrian E, Larena C, Martinez L. Comparison of Goldmann tonometry, pneumotonometry and the effect of central corneal thickness. *Arch Soc Esp Oftalmol* 2010;85(10):325-328.

20. Answer: d

Discussion: The Ishihara Pseudoisochromatic plates are unable to detect tritanopia. The test should be read in daylight or under artificial light resembling daylight. Prolonged exposure of the test colours to sunlight may cause fading of the colours and loss of accuracy.

Source:
1. James B, Benjamin L. 2007. *Ophthalmology Investigation and Examination Techniques*. Elsevier.

21. Answer: d

Discussion: Imaging in patients with clinically evident thyroid eye disease is often not indicated. CT is useful before decompressive surgery, while MRI may be more sensitive to detect optic nerve compression. Recti are often enlarged, with tendon sparing. Isolated rectus involvement is seen only in 5-6% of patients. Posterior globe tenting indicates stretching of the optic nerve caused by proptosis and is associated with a poorer visual outcome. Orbital fat on CT is hyperdense to normal patients and correlates positively with muscle volume.

Source:
1. Regensburg NI, Wiersinga WM, Berendschot TT, Saeed P, Mourits MP. Densities of orbital fat and extraocular muscles in graves orbitopathy patients and controls. *Ophthal Plast Reconstr Surg*. Jul-Aug 2011;27(4):236-40
2. Dalley RW, Robertson WD, Rootman J. Globe tenting: a sign of increased orbital tension. *AJNR*. 1989;10:181—6

22. Answer: b

Discussion: The following table reveals the MRI characteristics of different choroidal lesions.

Lesion	* T1	* T2	Contrast Enhancement
Choroidal Melanoma	Hyperintense	Hypointense	Marked
Choroidal Haemangioma	Hyperintense	Hyperintense	Marked
Choroidal Osteoma	Hyperintense	Hypointense	Central Core enhanced
Choroidal Metastasis	Hyperintense	Hypointense	Marked

* intensity in relation to vitreous

MRI presentations of choroidal melanomas, osteomas and metastases tend to be similar.
Clinical course and presentation. may aid in differentiating these radiologically similar lesions.
Osteomas are rare, ossifying choroidal lesions presenting in young women. They may be bilateral, and present as flat, yellow-red, peripapillary or macular lesions with slowly progressive pseudopod like edges, unlike the rapidly progressive, painful course of uveal metastases or the progressive elevation and presence of subretinal fluid associated with choroidal melanoma.

Source:
1. Depotter P, Shields J.A, Shields C.L, Rao V.M. Magnetic resonance imaging in choroidal osteoma. *Retina*. 11(2): 221-223
2. Baert AL, Sartor K. 2006. *Imaging of Orbital and Visual Pathway Pathology*. Springer-Verlag Berlin Heidelberg

23. Answer: b

Discussion: Bagolini striated glasses consist of parallel cylinders akin to Maddox rods. Their axes are placed perpendicular to each other, usually at 135 degrees over the right eye and 45 degrees over the left.

24. Answer: b

Discussion: Pharmacodynamics refers to the biological activity of the drug, including receptor affinity and subsequent cellular effects.

25. Answer: b

Discussion: Phenylephrine is a direct acting alpha 1 agonist. Phenylephrine mydriasis is often overcome by the light reflex during fundoscopy. The side effects of phenylephrine include hypertension, of particular concern in patients with orthostatic hypotension and patients on monoamine oxidase inhibitors or tricyclic antidepressants – which enhance the effects of phenylephrine. The use of phenylephrine should also be used with caution in infants as dose per body weight is larger.

26. Answer: c

Discussion: CCL has rapidly become the treatment of choice for progressive keratoconus in the paediatric patient who is at more risk of developing severe disease as compared to adult patients. Multiple studies have demonstrated rapid functional improvement and halting of progression of the disease post treatment.

Source:
1. Caporossi A, Mazotta C, Baiocchi C, et al. Riboflavin-UVA-induced corneal collagen cross linking in pediatric patients. *Cornea*. 31:227-231
2. Zotta PG. Corneal collagen cross-linking for progressive keratoconus in pediatric patients: a feasibility study. *J Refract Surg*. 2012 Nov;28(11):793-9.

27. Answer: d

Discussion: Antimetabolite use allows more control of post operative scarring and potential failure of glaucoma filtration surgery. During trabeculectomies, antimetabolites are applied beneath the conjunctival flap and to the scleral bed prior to entering the anterior chamber, avoiding conjunctival wound edges in order to minimize the risk of post operative wound leak.

As per NICE glaucoma guidelines, all patients with primary open angle glaucoma undergoing trabeculectomy should have augmentation with antimetabolites, unless contraindicated e.g pregnancy. Post operative needling is typically carried out with a total dose of 5 mg of 5-FU

Source:
1. National Institute if Clinical Excellence Guidelines: Glaucoma CG85

28. Answer: d

Discussion: The term *phacoanaphylaxis* is likely a misnomer as evidence of a classic Type 1 IgE mediated anaphylactic reaction does not currently exist.

Evidence for a Type 4 reaction after sensitization of the immune system to lens proteins exists.

Source:
1. Brinkman CJ, Broekhuyse RM. Cell mediated immunity in relation to cataract and cataract surgery. *Br J Ophthalmol*. May 1979;63(5):301-5

29. Answer: b

Discussion: OCP refers to a heterogenous group of disorders characterized by cicatricial blistering and inflammation predominantly involving the mucous membranes of the conjunctiva, upper airways and upper gastrointestinal tract. The skin is only occasionally involved.

A Type 2 hypersensitivity reaction predominates, involved epithelial basement membrane antigens include BPAG 1 and 2, Type 7 collagen, integrins and laminins. Patients are often in their 60s and women are twice as likely to be involved.

Source:
1. Chan LS, Ahmed AR, Anhalt GJ, Bernauer W, Cooper KD, Elder MJ, et al. *The first international consensus on mucous membrane pemphigoid: definition, diagnostic criteria, pathogenic factors, medical treatment and prognostic indicators*. Arch Dermatol. Mar 2002;138(3):370-9

30. Answer: c

Discussion: PUK refers to an inflammatory peripheral keratitis which may be associated with an underlying systemic collagen disorder/vasculitis. It is a predominantly Type 3 hypersensitivity reaction, with deposition of immune complexes at the limbus and subsequent inflammation, ulceration and thinning.

It is most commonly associated with rheumatoid arthritis, which may cause up to 34% of non - infectious PUK. It less commonly occurs in association with wegener's granulomatosis, polyarteritis nodosa, RP and systemic lupus. Systemic collagen disorders/vasculitides tend to occur more in women, accounting for the female predilection of PUK.

Source:
1. Eiferman RA, Carothers DJ, Yankeelov JA Jr. Peripheral rheumatoid ulceration and evidence for conjunctival collagenase production. *Am J Ophthalmol*. May 1979;87(5):703-9.
□□□Brown SI, Grayson M. Marginal furrows. A characteristic corneal lesion of rheumatoid arthritis. *Arch Ophthalmol*. May 1968;79(5):563-7

31. Answer: b

32. Answer: c

Discussion: Of the above disorders, sub retinal bleeding may occur secondary to choroidal neovascularisation in POHS, Choroidal Rupture and end stage Best's Disease. Bleeding in diabetic retinopathy tends to be limited to the retinal layers, the subhyaloid space as well as the vitreous.

33. Answer: c

Discussion: Achromatopsia refers to deficient function of one or all cones. Most patients have *complete achromatopsia* which refers to deficient function of all 3 cone populations. Patients tend to present at birth or shortly after with nystagmus. Eventual vision is often around 6/60 in patients with complete achromatopsia and may be as high as 6/24 in patients with incomplete disease. The fundus often appears normal.

Source:
1. Hartnett, ME. 2005. *Pediatric Retina*. Lipincott Williams and Wilkins.

34. Answer: d

Discussion: The drug causes of cystoid macular oedema include both systemic and topical drugs. Topical latanoprost and adrenaline are classic causes, while tamoxifen, oral niacin and intravenous interferon have all been implicated.

Source:
1. Hejny C, Sternberg P, Lawson DH et al. Retinopathy associated with high dose interferon alfa 2b therapy. *Am J Ophthalmol* 2001; 131:782-787
2. Jampol LM. Niacin maculopathy. *Ophthalmology* 1988;95:1704-1705
3. Alwitry A, Gardner I. Tamoxifen maculopathy. *Arch Ophthalmol* 2002;120:1402

35. Answer: d

Discussion: According to Royal College guidelines, intravitreal injections should be administered in a dedicated clean room or theatre dealing with non - infective cases. Injections should be 3.0-3.5 mm from the lmbus in pseudophakic patients, and 3.5-4.0 mm from the limbus in phakic patients. IOP checks are non mandatory but should be considered in patients with glaucoma. If the patient is NPL post injection, they should be placed supine, and an anterior chamber paracentesis is indicated.

Source:
1. Royal College of Ophthalmologists Guidelines for Intravitreal Injections Procedure 2009

36. Answer: d

Discussion: Salient points regarding consent are;
-No one is able to give consent on behalf of an adult patient with capacity.
-Should an adult patient be incapacitated, final responsibility of care lies with the attending medical team, taking into consideration the wishes of those closest to the patient, always choosing the option that is least restrictive.
- For children, consent for a procedure requires parental consent, unless the child is deemed capable of consenting (refer *Gillick Competence)*
- During counseling for a procedure, all significant complications must be mentioned, no matter how insignificant the risk; less significant complications should also be mentioned, should they occur frequently
- The doctor should not assume to know the amount of knowledge the patient is likely to want to hear regarding the procedure, but should offer as much information as the patient needs in order to allow them to make an informed decision

Source:
1. General Medical Council Guidelines on *Consent: patients and doctors making decisions together* 2013

37. Answer: c

Discussion: Sensitivity is the proportion of patients testing positive divided by number of patients with the disease (True positives + false negatives). The false negative rate = 1-sensitivity. Higher sensitivity is desirable for screening tests.
Specificity is the proportion of patients testing negative who do not have the disease. False positive rate = 1-specificity. A higher specificity is desirable for diagnostic or confirmatory procedures.

38. Answer: a

Discussion: CSD is caused by the gram negative bacillus *Bartonella henselae*. Inoculation typically occurs due to the scratch or bite of a kitten. Regional lymphadenopathy occurs within 2 weeks from inoculation and the disease resolves spontaneously within 2-4 months. Rarely, systemic manifestations may occur including oculoglandular syndrome (unilateral granulomatous conjunctivitis), encephalitis, neuroretinitis, osteomyelitis, arthralgia and erythema nodosum.

Source:
1. Regnery R, Tappero J. Unraveling mysteries associated with catscratch disease, bacillary angiomatosis and related syndromes. *Emerg Infect Dis.* Jan-Mar 1995;1(1):16-21
2. Carithers HA. Catscratch disease. An overview based on a study of 1200 patients. *Am J Dis Child.* Nov 1985;139(11):1124-33

39. Answer: a

Discussion: There currently is no data to support steroid use in NA-AION. Optic nerve fenestration as per the findings of the Ischemic Optic Neuropathy Decompression Trial confers no benefit. Once an eye has been involved, second episodes do not occur, and thus vision tends to remain stable. However there is a risk of contralateral eye involvement in around 19% of patients.

Source:
1. The Ischemic Optic Neuropathy Decompression Trial Research Group. Optic nerve decompression surgery for non-arteritic anterior ischemic optic neuropathy (NAION) is not effective and may be harmful. The Ischemic Optic Neuropathy Decompression Trial Research Group. *JAMA.* Feb 22 1995;273(8):625-32.

40. Answer: c

Discussion: LHON has a mitochondrial inheritance pattern (maternally inherited). Point mutations involving respiratory chain enzymes are responsible, the most common being 11778, comprising 95% of cases. Typical presentation is sudden, sequential painless blurring of vision in otherwise healthy young adult males.

41. Answer: c

Discussion: Neuromyelitis Optica or Devic disease, is an idiopathic, demyelinating disorder which typically affects the optic nerves and spinal cord. IgG antibodies to aquaporin 4, the main water channel protein responsible for water homeostasis within the central nervous system, are diagnostic. Histologically, there is perivascular deposition of IgG and complement as well as antibody directed granulocyte infiltration. The disease frequently relapses, and because optimal treatment differs from that for multiple sclerosis, early recognition and diagnosis is critical.

Source:
1. Wingerchuk DM. The spectrum of neuromyelitis optica. *Lancet Neurol* 2007; 6: 805–15

42. Answer: a

Discussion: SCC is much less common than basal cell carcinomas (BCC), which make up 90% of all head and neck cutaneous malignancies. When involving the eyelids, SCC tends to involve the lower lids. Risk factors are many and include ultraviolet (UV) radiation; immunosuppression; use of tobacco or alcohol; age; familial or genetic predisposition; nutritional status; chronic irritation; and exposure to industrial products or heavy metals, viruses, or ionizing radiation.

Poor prognostic factors include tumor size > 2cm, fat invasion, undifferentiated tumor, perineural infiltration (which may cause paraesthesia, tingling, or muscular weakness) and ear or temple location.

43. Answer: a

Discussion: Marcus Gunn Jaw Wink is a congenital disorder of aberrant innervation, linking innervation of the pterygoid muscles and the ipsilateral levator muscle, which causes elevation of the ptotic lid during chewing, smiling, or movement of the jaw from side to side.

44. Answer: b

Discussion: Cyclic Esotropia is a rare condition presenting in childhood.
It is characterised by cycles of esotropia with suppression followed by a period of straight eyes with binocular single vision. The duration of the cycles is remarkably constant.
The angles are similar at distance and at near, there is usually no significant refractive error and because the deviations are not persistent, amblyopia is rare.
Decompensating esophorias may simulate cyclic esotropia, but a good history will often distinguish the two.

Source:
1. MacEwen C, Gregson R. 2003. *Manual of Strabismus Surgery*. London. Elsevier Limited.
2. Costenbader FD, Mousel DK. Cyclic Esotropia. *Arch Ophthalmol*. 1964;71(2):180-181.

45. Answer: d

Discussion: Presentation of retained copper may be divided into 2 main presentations – an acute, early, fulminant endophthalmitis-like picture or a chronic presentation, characterised by reducing visual acuity, Keyser Fleischer rings, glaucoma, greenish iris discolouration, sunflower cataracts and a flat ERG.

Management of a copper foreign body therefore often depends on timing of presentation – in patients presenting early, removal is indicated given the risk of endophthalmitis or inflammation. In patients with a delayed presentation of an otherwise uncomplicated foreign body, the risk of chronic chalcosis should be weighed against the risk of removal. The patient can be monitored with serial ERG and visual acuity.

Source:
1. Neubauer H (1979) The Montgomery Lecture, 1979. Ocular metallosis. *Trans Ophthalmol Soc* UK 99: 502–510
2. Budde WM, Junemann A (1998) Chalcosis oculi. *Klin Monatsbl Augenheilk* 212: 184–185

46. Answer: d

Discussion: White without pressure is predominantly seen in Asians and Afro-Carribean patients. It presents as a whitish sheen of the retina akin to that caused by indentation. Choroidal markings are usually absent, and it may be confused with shallow detachments.

47. Answer: a

Discussion: As per RCOphth Retinal Vein Occlusion guidelines, options for a patient with branch retinal vein occlusion presenting within 3 months with macular oedema, acuity of 6/12 or less and foveal thickness of >250μm include intravitreal dexamethasone or ranibizumab.
Intravitreal dexamethasone implant is the licensed option.

Source:
1. Royal College of Ophthalmologists Interim Guidelines for Retinal Vein Occlusion 2010

48. Answer: d

Discussion: Fresnel prisms, by virtue of easy changing and much lower cost, should be the option of choice for temporary management of acquired diplopia that is expected to improve. They tend to cause aberrations that worsen as their power increases, and thus, Fresnel prisms of 12 dioptres or more are inferior to ground in prisms of similar power. They should also only be fitted to one eye, usually the non - dominant eye, in order to minimise the degradation of visual acuity.

Source:
1. Veronneau-Troutman S. Fresnel prisms and their effects on visual acuity and binocularity. *Trans Am Ophthalmol Soc* 1978; 76:610–653

49. Answer: d

Discussion: Valganciclovir is a prodrug of ganciclovir which is available orally. It has excellent bioavailability and has largely superseded the use of intravenous ganciclovir for induction and maintenance therapy. Its advantages lie in the fact that dosing is more convenient as compared to that of ganciclovir which requires twice daily intravenous dosing for induction. As oral ganciclovir has a bioavailability of less than 10%, this is not a reasonable choice for induction.
Recent literature suggests that oral ganciclovir is non-inferior to intravenous ganciclovir in treatment of CMV retinitis in transplant patients. It does however, carries a risk of neutropenia similar to that of ganciclovir.

Source:
1. Asberg, A. Oral valganciclovir is noninferior to intravenous ganciclovir for the treatment of cytomegalovirus disease in solid organ transplant recipients. *Am J Transplant.* 2007 Sep;7(9):2106-13.
2. Altaweel, M. CMV Retinitis. *www.medscape.emedicine.com*

50. Answer: c

Discussion: Cicatricial entropion can be managed surgically according to the severity of the scarring.
In mild to moderate scarring, anterior lamella repositioning is often the procedure of choice. In more severe scarring with lid retraction, a tarsal fracture and posterior lamellar graft may be needed.

Source:
1. Tyers AG, Collin JRO. 2008. *Colour Atlas of Ophthalmic Plastic Surgery.* 3rd Ed. Elsevier

51. Answer: d

Discussion: All patients with a CCT of <555μm or an IOP of > 32mmHg should be treated with a prostaglandin analogue.
Patients with a CCT of 555-590μm should be treated with a beta blocker only if their IOP >25mmHg.
Patients with a CCT of >590μm should be treated if their IOP >32mmHg.

Source:
1. National Institute of Clinical Excellence Guidelines CG85: Glaucoma

52. Answer: b

Discussion: According to GMC guidelines on confidentiality, disclosure of information to a sexual contact of a patient with a communicable sexual disease is allowed if the contact is believed to be at risk and the patient cannot be persuaded to inform them. In the above scenario, as Mr. A has not refused outright to inform his wife and GP regarding his HIV status, he should be persuaded to and supported throughout the process.

Should he refuse, it is reasonable to disclose information without his consent as the safety of his wife is at stake – he must however be informed of your decision to disclose the information prior to the disclosure.

Source:
1. General Medical Council Guidelines on Confidentiality October 2009 : *disclosing information about serious communicable diseases, paragraph 10*

53. Answer: b

Discussion: The odds ratio is calculated as the odds of having the disease in the exposed/intervention group divided by the odds of having the disease in the unexposed/control group, which in this case would be

$$= \frac{30/10}{25/25} = 3/1 = 3$$

54. Answer: b

Discussion: While both anterior and posterior subcapsular cataracts are seen in atopic dermatitis, anterior subcapsular cataracts are more common. Reduced function of superoxide dismutase has been reported, suggesting cataract formation is secondary to chronic inflammation.

Source:
1. Bair B, Dodd J, Heidelberg K, Krach K. Cataracts in atopic dermatitis: a case presentation and review of the literature. *Arch Dermatol* 2011 May;147(5):585-8.

55. Answer: b

Discussion: Technically, by placing the haptics in the sulcus and capturing the optic into the bag behind the capsulorhexis, the original lens may be used. Limited data supports the technique, demonstrating post - operative refraction within 1 dioptre of the targeted refraction.
However, a sulcus implant without an optic capture is usually 0.5 dioptres less than originally planned.

Source:
1. Jones JJ, Oetting TA, Rogers GM, Jin GJ. Reverse optic capture of the single-piece acrylic intraocular lens in eyes with posterior capsule rupture. *Ophthalmic Surg Lasers Imaging.* 2012 Nov-Dec;43(6):480-8.

56. Answer: b

Discussion: Less viscous materials coat surfaces more evenly and effectively and hence are better at protecting corneal endothelium.

57. Answer: a

Discussion: Tilted discs are common, bilateral, often asymmetrical congenital anomalies of the optic nerve caused by oblique entry of the nerve into the globe. This causes the typical inferonasal orientation of the disc. They are associated with myopia. Associated inferonasal peripapillary atrophy may cause the presence of bilateral supero – temporal field defects which are static. In contrast to chiasmal lesions, the defects do not respect the vertical midline.

58. Answer: c

Discussion: Filamentary keratitis is most often seen in the context of keratoconjunctivitis sicca. Mucous filaments attached to punctate erosions may cause significant foreign body sensation and stinging. The mainstay of treatment is to manage the underlying disorder. Preservative free lubrication, 5% hypertonic saline and N acetylcysteine as a mucolytic agent are all useful.

Source:
1. Albietz J, Sanfilippo P, Troutbeck R, Lenton LM. Management of filamentary keratitis associated with aqueous-deficient dry eye. *Optom Vis Sci.* 2003 Jun;80(6):420-30.

59. Answer: b

Discussion: OAL differs greatly in terms of natural history and treatment as compared to primary intraocular lymphoma (PIOL), which is a subset of CNS lymphoma. A recent systematic review reported that OAL comprises of predominantly low-grade B cell lymphomas, whereas PIOL is predominantly high grade, and in contrast to OAL, associated with HIV.
Chemotherapy is the mainstay of treatment for PIOL, whereas radiotherapy is the mainstay of treatment for OAL.

Source:
1. Decaudin D, de-Cremoux P, Vincent-Salomon A, Dendale R, Lumbroso-Le Rouic L. Ocular adnexal lymphoma: a review of clinicopathologic features and treatment options. *Blood.* 2006 108: 1451-1460

60. Answer: b

Discussion: The abovementioned scenario suggests a chiasmal lesion as evidenced by bilateral bow-tie atrophy. Lesions compressing the centre of the chiasm involve the crossing fibers from the nasal retina, which tend to end in the nasal or temporal rims of the optic disc, causing a horizontal band of pallor. In contrast, post geniculate lesions do not give rise to disc pallor, effectively excluding tract and internal capsular lesions as a cause.

Source:
1. Newman NJ, Miller NR, Biousse V. 2008. *Walsh and Hoyt's Clinical Neuro-Ophthalmology: The Essentials.* 2nd Ed. Lippincott Williams & Wilkins

61. Answer: c

Discussion: The differential diagnoses for intermittent proptosis include orbital mucoceles, (which may increase in size with inflammation) lymphangiomas, haemangiomas and varices.
Orbital cavernous haemangiomas do not present with intermittent proptosis.

Source:
1. Kanski JJ. 2010. *Signs in Ophthalmology: Causes and Differential Diagnosis.* Elsevier

62. Answer: c

Discussion: Congenital CMV is the most common intrauterine infection, estimated to account for up to 2.4% of all live births. The risk of intrauterine infection is highest with primary maternal infection. 90% of infected neonates are asymptomatic during the neonatal period. The classic ocular finding is a chorioretinitis, occurring in up to 20% of infected neonates. Other ocular findings include anophthalmia, corneal opacities, cataracts and optic nerve hypoplasia.

Intravenous ganciclovir and foscarnet are unable to completely eliminate the virus – viruria returns to pre-treatment levels after cessation of the drugs.

Source:
1. Bartlett JG: *The Johns Hopkins Hospital guide to medical care of patients with HIV infection.* Baltimore, Williams and Wilkins, 1997
2. Coats DK, Demmler GJ, Paysse EA, et al: Ophthalmologic findings in children with congenital cytomegalovirus infection. *J AAPOS* 4:110--6, 2000
3. Dobbins JG, Stewart JA, Demmler GJ: Surveillance of congenital cytomegalovirus disease, 1990--1991. Collaborating Registry Group. *MMWR CDC Surveill Summ* 41:35--9, 1992

63. Answer: d

Discussion: 40% of patients with orbital atypical lymphoid hyperplasia will develop systemic lymphoma within 5 years. Many authorities thus recommend 6 monthly follow-ups with systemic investigations to detect systemic disease.

Source:
1. Friedman NJ, Kaiser PK. 2007. *Essentials of Ophthalmology.* Elsevier Inc.

64. Answer: d

Discussion: 85% of Duane Syndrome cases are Type 1. Type 1 preferentially affects girls (60%) and the left eye (60%). It is unilateral in up to 80% of cases.

Source:
1. Huber A. Electrophysiology of the retraction syndromes. *Br J Ophthalmol.* Mar 1974;58(3):293-300.

65. Answer: d

Discussion: Typical findings in FAS include short horizontal fissures, telecanthus, ptosis, esotropia and myopia.

A diverse range of other features that occur less commonly have also been described, ranging from microphthalmos, cataract, glaucoma and optic nerve hypoplasia (which is the most common posterior segment abnormality associated with FAS).

Source:
1. Strömland K, Pinazo-Durán MD. Ophthalmic Involvement in the Fetal Alcohol Syndrome. Clinical and Animal Model Studies. *Alcohol & Alcoholism* Vol. 37, No. 1, pp. 2–8, 2002

66. Answer: b

Discussion: Botulinum toxin is therapeutic in primary and secondary neurogenic strabismus in correcting the angle of deviation to allow restoration of, or to allow the development of binocular single vision.

It is very much less effective in correcting the angle of deviation in restrictive strabismus when recession of the fibrotic muscle is more effective.

Source:
1. Rosenbaum AL, Santiago AP, Hunter D. 1st Edition. *Clinical Strabismus Management: Principles and Surgical Techniques.* Philadelphia, Pennsylvania. W.B Saunders Company.

67. Answer: b

Discussion: ERG may be a useful modality to detect early signs of toxicity in patients with an otherwise uncomplicated intraocular iron foreign body. Typical early changes include a supranormal a wave and a normal or supranormal 'b' wave, followed by gradual loss of the b wave and eventual loss of the ERG.

Source:
1. Weiss MJ. Ocular siderosis. Diagnosis and management. *Retina.* 1997;17(2):105-8.
2. Karpe G. Early diagnosis of siderosis retinae by the use of electroretinography. *Doc Ophthalmol* 01/1948; 2(1):277-296.

68. Answer: c

Discussion: Juvenile retinoschisis is an X linked, recessive maculopathy that affects boys almost exclusively. Female carriers are not affected in any way. The defective gene is the XLRS1 gene on the short arm of chromosome X. The most constant finding is an early stellate or cystoid maculopathy which is present in almost every case, and eventually leads to a macular scar.
Visual acuity deteriorates rapidly during the first 5 years of life, and then the rate of deterioration slows. Visual acuity has been reported to be 6/18 at the age of 20 and 6/60 at the age of 60.
Juvenile retinoschisis is not associated with systemic findings, and prophylactic treatment to prevent detachments is not advisable.

Source:
1. Wright KW, Spiegel PH. 2003. *Pediatric Ophthalmology and Strabismus.* 2nd Ed. New York. Springer-Verlag.

69. Answer: b

Discussion: Causes of band keratopathy may either be *ocular:* glaucoma/uveitis/silicone oil keratopathy or *systemic:* chronic renal impairment, vitamin d toxicity, hyperparathyroidism, hypercalcemia and ichthyosis.

70. Answer: b

Discussion: Phacolytic glaucoma is a secondary open angle glaucoma caused by leakage of soluble lens protein into the anterior chamber through a macroscopically intact anterior capsule. It tends to happen in mature or hypermature cataracts, and rarely, in immature cataracts.
Incidence is higher in underdeveloped countries and there is no gender bias.

71. Answer: b

Discussion: Medical treatment should be first line options for aqueous misdirection syndrome. Cycloplegics allow posterior movement of the lens iris diaphragm and allows any present cilio-lenticular block to be broken.
Pilocarpine is contraindicated as miotics cause relaxation of the ciliary muscle and forward movement of the lens iris diaphragm.
Hyperosmolar agents are effective as they work by shrinking vitreous volume. Beta and alpha - blockers work by reducing aqueous production by the ciliary processes.

72. Answer: a

Discussion: Lattice Dystrophy Type 1 is associated with mutations of the BIGH3 gene that codes for the basement membrane protein keratoepithelin. Type 2 is associated with mutations in the gene coding for the cytoskeletal protein gelsolin.
Both types present with recurrent corneal erosions (Type 1 in the 1st decade of life, and Type 2 in the 3rd decade) Type 3 tends to present with painless blurring of vision in the 4th to 6th decades.
Systemic amyloidosis with peripheral neuropathies, cardiac or renal failure is associated with Type 2.
Recurrences are common after keratoplasty.

Source:
1. Mashima Y, Yamamoto S, Inoue Y et al. Association of autosomal dominantly inherited corneal dystrophies with BIGH3 gene mutations in Japan. *Am J Ophthalmol.* Oct 2000;130(4):516-7

73. Answer: d

Discussion: causes of Conjunctival pigmentation can be divided into *congenital* or *acquired* lesions that may either be benign or malignant. *Congenital* pigmented lesions include
Benign Conjunctival Melanosis (freely moving, patchy, asymmetric pigmentation of the conjunctiva usually denser along limbus, involving dark skinned individuals)
Congenital Ocular Melanosis (deep, slate grey hyperpigmentation involving the episclera/uvea, non movable, which may be associated with iris hyperpigmentation and risk of uveal melanoma)
Conjunctival Nevi (Nevi are well defined, single, pigmented lesions, most often located at the nasal or temporal limbus. They may exhibit cysts and feeder vessels.)

In contrast, **Melanomas** and **Melanocytomas** are acquired conditions.

Source:
1. Kanski JJ. 2007. *Clinical Ophthalmology: A Systematic Approach.* 6th Ed. Elsevier.

74. Answer: d

Discussion: Choroideremia is X linked recessive and usually presents in males. Female carriers are usually asymptomatic.

75. Answer: b

Discussion: As per Royal College and Department of Health guidelines, all healthcare workers, at the commencement of work with the NHS should be offered screening for HIV and Hepatitis B and C.
In addition, all surgeons performing exposure prone procedures MUST be HIV negative, Hepatitis B negative (or if positive, e antigen negative with a low viral load) and Hepatitis C negative (or if positive, C RNA negative). There are no restrictions placed on surgeons performing only none exposure prone procedures.

Source:
1. Royal College Guidelines on Bloor Borne Viral Infections September 2008

76. Answer: a

Discussion: The more common disorders associated with angioid streaks include pseudoxanthoma elasticum, sickle cell disease and paget's disease.
Less common aetiologies include gastrointestinal ulceration, breast malignancies, hypertension, diabetes, rheumatoid spondylitis and cardiac disease.

Source:
1. Abusamak, M. *Angioid Streaks*. emedicine.medscape.com

77. Answer: c

Discussion: Active disease presents either with multifocal or diffuse stromal inflammation usually affecting the deeper layers of the stroma without significant involvement of the endothelium or epithelium. New vessels may grow into the stroma, causing a pink salmon patch by reflection of light off the deep vessels that is scattered by overlying stromal oedema. The vessels may rupture, causing haemorrhage.
Eventually the inflammation regresses, leaving corneal thinning and scarring with ghost vessels.

78. Answer: d

Discussion: Higher total dose, higher dose per fraction as well as concurrent chemotherapy and diabetes are all risk factors of developing radiation retinopathy. Fewer fractions for a given total dose equates to larger doses per fraction.

Source:
1. Parsons JT, Bova FJ, Fitzgerald CR et al. Radiation retinopathy after external beam irradiation: analysis of time-dose factors. *Int J Radiat Oncol Biol Phys* 1994; 30:765-773

79. Answer: b

Answer: The scenario above suggests a left sixth nerve palsy in association with contralateral hemiparesis. The presentation of 6th nerve palsy can be divided into isolated 6th nerve palsies or complex where the nerve palsy is associated with other findings. This scenario suggests Raymond's syndrome, where the lesion in the ventral pons affects both the descending corticospinal tract and the fasciculus of the 6th nerve passing through the tract.

Source:
1. Satake M, Kira J, Yamada T, Kobayashi T. Raymond syndrome (alternating abducent hemiplegia) caused by a small haematoma at the medial pontomedullary junction. *J Neurol Neurosurg Psychiatry.* Feb 1995; 58(2): 261.

80. Answer: b

Discussion: Terrien's preferentially affects men, and is characterized by progressive, bilateral (often asymmetrical) superior corneal thinning. There is slow and progressive flattening of the central vertical meridian leading to against the rule astigmatism. It is often painless, with a lipid line posterior to the area of thinning.

Source:
1. Brightbill FS, McDonnell PJ, McGhee CNJ, Farjo AA, Serdarevic O. 2009. *Corneal Surgery: Theory, Technique and Tissue.* 4th Ed. Elsevier.

81. Answer: b

Discussion: Choroidal melanomas may spread in 3 ways; haematogenous (which is most common, and commonly to the liver), transclerally into the orbit via emissary channels, and rarely, via the optic nerve (which occurs only with peripapillary tumors) Lymphatic spread is not possible as the eye lacks lymphatic channels.

82. Answer: c

Discussion: Intracranial aneurysms may carry a mortality rate as high as 65% in cases of rupture with subsequent SAH. Non-ruptured aneurysms may cause compression effects that may differ according to location of the aneurysm.
Motor neuropathies including isolated 3rd nerve, trochlear or 6th nerve palsies may be the first sign on an otherwise asymptomatic aneurysm. Combined oculomotor palsies may be caused by aneurysms of the basilar artery.
Aneurysms of the anterior cerebral or anterior communicating artery may cause unilateral visual loss due to optic nerve compression. Intracavernous carotid artery aneurysms may compress the temporal aspect of the tract with a resulting ipsilateral nasal field loss.
MCA aneurysms may cause a homonymous hemianopia due to compression of the optic radiations.

Source:
1. Newman NJ, Miller NR, Biousse V. 2008. *Walsh and Hoyt's Clinical Neuro-Ophthalmology: The Essentials.* 2nd Ed. Lippincott Williams & Wilkins

83. Answer: c

Discussion: Causes of nyctalopia in childhood can be divided into *stationary* and *progressive* causes. Stationery causes include the heterogenous group of disorders collectively described as Congenital Stationary Night Blindness (CSNB). This group includes X linked, Autosomal Recessive and Autosomal Dominant CSNB, Oguchi's Disease and Fundus Albipunctatus.
Causes of progressive nyctalopia in childhood include retinitis pigmentosa and its variants, including Refsum's Disease.
Stargardt's Disease and Fundus Flavimaculatus are macular dystrophies and present with progressively worsening visual acuity.

Source:
1. Brodsky MC. 2010. *Pediatric Neuro-Ophthalmology.* 2nd Edition. Springer

84. Answer: c

Discussion: Surgical decompression should be offered to patients with sight threatening thyroid eye disease who are unable or who are likely to be unable to tolerate intravenous methylprednisolone.

Source:
1. Bartalena L, Baldeschi L, Dickinson AJ, Eckstein A, Kendall-Taylor P, Marcocci C. et al. Consensus Statement of the European Groupon Graves' Orbitopathy (EUGOGO) on Management of Graves' Orbitopathy· *Thyroid.* Volume 18, Number 3, 2008

85. Answer: b

Discussion: Mortality in non-treated individuals infected with HIV is 90%, the vast majority of which occur within 2 years of developing full-blown AIDS. The mortality of HIV/AIDS is related to the mode of transmission, being highest in patients infected via needle sharing. The advent of HAART in 1995-1997 has dramatically increased survival rates and the risk of opportunistic infection. HAART has no effect on the rates of HIV/AIDS associated lymphoma. Current WHO guidelines strongly recommend initiating HAART when CD4 counts are <350/mm3. NNRTIs are unfortunately, non-effective against HIV-2.

Source:
1.Bennett N.J, Gilroy S.A. *HIV Disease*. Emedicine.medscape.com
2. WHO HAART guidelines 2013 http://www.natap.org/2012/HIV/033012_01.htm

86. Answer: d

Discussion: Apraxia of Lid Opening (ALO) is characterized by an inability to voluntarily initiate the opening of closed eyelids. It is thought to be related to supranuclear dysfunction that causes inhibition of the levator palpebrae as well as involuntary contraction of the obicularis oculi.
Causes include Parkinson's disease, secondary Parkinsonian syndromes (including PSP) as well as Huntington's disease.
Gullian Barre is an acute, inflammatory disorder that tends to cause lagophthalmos due to involvement of the facial nerve.

Source:
1. Lepore FE, Duvoisin RC. "Apraxia" of eyelid opening: an involuntary levator inhibition. *Neurology*. Mar 1985;35(3):423-7.
2. Golbe LI, Davis PH, Lepore FE. Eyelid movement abnormalities in progressive supranuclear palsy. *Mov Disord*. 1989;4(4):297-302
3. Goldstein JE, Cogan DG. Apraxia of Lid Opening. *Arch Ophthalmol*. Feb 1965;73:155-9.

87. Answer: a

Discussion: Drugs likely to cause nystagmus as a side effect include anticonvulsants and barbiturates. Gabapentin is often used to treat nystagmus.

Source:
1. Shery T, Proudlock FA, Sarvananthan N, Mclean RJ, Gottlob I. The effects of gabapentin and memantine in acquired and congenital nystagmus: a retrospective study. *Br J Ophthalmol*. 2006 July; 90(7): 839–843.

88. Answer: d

Discussion: This is likely Horner syndrome secondary to a traumatic dissection of the thoracic aorta. Ultrasonography is less reliable in diagnosing aortic dissection as compared to a CT angiogram. Intravenous catheter angiography remains the gold standard.

Source:
1. Bardorf CM. Horner Syndrome. *emedicine.medscape.com*

89. Answer: c

Discussion: AS tends to affect young men. However, recent data seems to suggest women are equally affected, but tend to present with less severe disease. 90% of individuals with AS are HLA B27+, and acute anterior uveitis occurs in 20-30% of patients.

Source:
1. Hart FD, Robinson KC. Ankylosing spondylitis in women. *Ann Rheum Dis* 1959;18:15-23.

90. Answer: b

Discussion: One of the ways to minimize publication bias is to ensure all studies are government registered, followed through, and results published, irrespective of outcome.

Source:
1. Wang D, Bakhai A. 2006. *Clinical Trials: A Practical Guide to Design, Analysis, and Reporting*. Remedica.

Paper 3

1. Which of the following trisomy syndromes with ocular significance is LEAST common?

a. Edwards Syndrome
b. Down Syndrome
c. Patau Syndrome
d. Warkany Syndrome 2

2. Which of the following is LEAST likely to be a risk factor for Creutzfeld-Jakob Disease (CJD) transmission?

a. Multiple blood transfusions
b. Previous neurosurgery or spinal surgery
c. Somatotrophin therapy
d. Corneal Transplantation

3. Which of the following statements regarding myopia in children is MOST likely to be false?

a. Usually develops between 6 and 14 years of age
b. Low birth weight is a protective factor
c. There is often a family history
d. Associated with close work and scholastic achievement

4. SLE is an example of

a. Type 1 Hypersensitivity
b. Type 2 Hypersensitivity
c. Type 3 Hypersensitivity
d. Type 4 Hypersensitivity

5. Which of the following statements regarding the genetics of retinoblastoma is MOST likely to be true?

a. The involved oncogene is a mutated Rb1 gene on the short arm of Chromosome 13
b. A germ line mutation gives rise to hereditary retinoblastoma
c. Hereditary retinoblastoma never arises from parental mosaicism
d. A child who acquires mutations to both Rb1 genes in a single retinal cell is likely to transmit an increased risk of developing retinoblastoma to his or her children

6. Which of the following is NOT an obligate anaerobic bacterium?

a. *Clostridia*
b. *Actinomyces*
c. *Bacteroides*
d. *Pseudomonas*

7. The following are expected complications during cataract surgery for mature, brunescent cataracts EXCEPT?

a. Poor visualization of the anterior capsule during capsulorhexis
b. Increased risk of radial tears during capsulorhexis
c. Small pupils
d. Zonular weakness

8. Which of the following iron lines is seen in patients with pterygium?

a. Fleischer
b. Stocker
c. Ferry's
d. Hudson-Stahli

9. The following are true regarding glaucoma in iridocorneal endothelial (ICE) syndrome EXCEPT?

a. Glaucoma is often worst in patients with predominantly corneal features
b. It is predominantly due to secondary angle closure
c. Augmented trabeculectomy is an acceptable surgical method of management
d. Success rates for augmented trabeculectomies are in the region of 60-70% at 1 year

10. What is the proportion of patients who will develop retinal neovascularization in a patient with an ischaemic branch retinal vein occlusion (BRVO)?

a. 16%
b. 26%
c. 36%
d. 46%

11. What is the MOST common field defect seen in ethambutol toxicity?

a. Hemianopic
b. Enlarged blind spot
c. Centrocaecal
d. Constricted visual field

12. Regarding malignant melanoma of the eyelids, which statement is MOST likely to be false?

a. The incidence of melanoma in white skinned individuals increases with decreasing latitude
b. May be associated with use of tanning beds
c. The subtypes of cutaneous melanoma correlate strongly with overall prognosis
d. The most common subtype is superficial spreading melanoma

13. Which of the following mucopolysaccharidoses (MPS) does not present with cloudy corneas?

a. Hurler
b. Hunter
c. Maroteaux–Lamy
d. Sly

14. Regarding the following scenario, which is the MOST likely diagnosis?

A 6-year-old girl attends your Strabismus clinic. She wears thick convex glasses. Her eyes are straight at distance and at near with her glasses on.
With her glasses off, she has an esophoria at distance and an alternating esodeviation at near. Ocular motility is full.

a. Partially Accommodative Esotropia
b. Fully Accommodative Esotropia
c. Convergence Excess Esotropia
d. Congenital Esotropia

15. The following are risk factors for traumatic endophthalmitis in a patient with an open globe injury EXCEPT?

a. Surgical closure delayed for 24 hours
b. Soil contamination
c. Shrapnel injury
d. Retained intraocular foreign body (IOFB)

16. The following statements are true of Familial Exudative Vitreoretinopathy (FEVR) EXCEPT?

a. Is autosomal dominant
b. Refraction is useful in differentiating from retinopathy of prematurity (ROP)
c. The predominant histopathological abnormality is poor perfusion of the neurosensory retina
d. Is unlike Coat's disease in that exudative detachments are often associated with a tractional component

17. Which of the following statements regarding specular microscopy is INCORRECT?

a. The sampled area becomes representative of the entire cornea
b. Cell counts in a patient with decompensated fuchs endothelial dystrophy would be expected to be less than 2000
c. Images are formed by light reflecting back from the cornea-aqeous interface and captured as a microphotograph
d. Significant pleiomorphism and polymegethism may be signs of aged related changes

18. Which of the following is NOT a contraindication for a contrasted computed tomographic (CT) scan?

a. Pregnancy
b. Renal impairment
c. Thyrotoxicosis
d. Shellfish Allergy

19. Which of the following is MOST likely to be the best option for disinfecting reusable tonometer tips?

a. Dry wiping for 10 seconds
b. 70% alcohol wipes for 10 seconds
c. 3% hydrogen peroxide soak for 1 minute
d. 3% hydrogen peroxide soak for 3 minutes

20. Which of the following statements about the Ishihara Pseudoisochromatic plates is LEAST likely to be false?

a. Can be read by a patient with 6/96 visual acuity
b. Utilizes difference in luminance between different dots to detect different numbers from surrounding dots
c. Should be read out of doors under direct sunlight
d. Results are markedly abnormal in patients with previous history of acute demyelinating optic neuritis

21. Which of the following statements is CORRECT regarding haematologic and immunological studies in thyroid eye disease?

a. Patients who are positive for thyroid stimulating hormone (TSH) antibodies always have symptoms of thyroid eye disease (TED)
b. TSH antibodies are seen in 20% of patients with TED
c. Antibodies to collagen XIII are highly indicative of active TED
d. A microcytic anaemia may be present

22. According to the following investigations, which is the MOST likely diagnosis?

A 40-year-old lady presents with unilateral blurring of vision. She has a well-demarcated orange fundal lesion next to the optic disc in her right eye. There is macular oedema. Ultrasound reveals a highly reflective anterior border with high internal reflectivity. There is no orbital shadowing. Fluorescein angiography (FA) reveals pre-arterial hyperfluorescence that increases throughout the angiogram with diffuse, late leakage.

a. Choroidal Melanoma
b. Choroidal Haemangioma
c. Melanocytoma
d. Uveal Lymphoma

23. Regarding tests for stereopsis, which of the following statements is LEAST likely to be true?

a. Random dot tests are the most effective for detection of high grade stereopsis
b. The Frisby stereo test does not require the use of spectacles
c. The Titmus Fly test allows stereopsis detection of up to 300 seconds of arc
d. The Lang test may be used to detect stereopsis in infants and very young children.

24. The following are true about topically administered drugs EXCEPT?

a. The movement of drug into the eye is largely through passive diffusion
b. The contact duration with the cornea is critical
c. Perilimbal conjunctiva offers an effective alternative to trans-corneal diffusion
d. It is advantageous to have a topically administered drug exist in a purely lipophilic form

25. Choose the MOST appropriate drug for the following description:

Transcription factor inhibitor that inhibits production of interleukin-2. Known to cause nephrotoxicity in 30% of patients.

a. Cyclophosphamide
b. Cyclosporin
c. Methotrexate
d. Mycophenolate

26. Which of the following statements about apraclonidine is FALSE?

a. Not suitable for long term treatment of glaucoma
b. May cause blepharoconjunctivitis in up to 40% of patients
c. Improves trabecular outflow
d. Has a superior side effect profile to brimonidine

27. Which of the following regarding Intraoperative Floppy Iris Syndrome (IFIS) is INCORRECT?

a. Associated with alpha-1-antagonists, particularly Tamsulosin (Flomax®)
b. Characterised by pre-operative small pupil
c. Increased propensity for iris prolapse through side port incisions
d. Stopping Tamsulosin reduces the risk of IFIS

28. Which of the following statements is LIKELIEST to be correct regarding chlamydial inclusion conjunctivitis?

a. Is caused by serotypes A-C
b. It is usually transmitted via aerosol spread
c. First line treatment for a 25-year-old lady should be oral doxycycline 100 mg BD
d. Non-tender lymphadenopathy is a typical finding

29. Which of the following statements regarding the features of Stevens Johnson Syndrome (SJS) and Ocular Cicatricial Pemphigoid (OCP) is LIKELIEST to be false?

a. SJS tends to occur in younger patients
b. Cutaneous involvement in SJS is the norm, while rare in OCP
c. Systemic immunosuppression is the mainstay of therapy in OCP while SJS tends to be self-limiting
d. OCP tends to affect men more than women

30. The following statements are true regarding deposition keratopathies EXCEPT?

a. Wilson's disease is associated with a brownish, peripheral ring at the level of Descemet's membrane
b. Vortex Keratopathy is usually innocuous
c. Infectious crystalline keratopathy is usually caused by *Staphylococcus epidermidis*
d. Amiodarone is a known cause of Vortex Keratopathy

31. Regarding pigment dispersion syndrome (PDS), which of the following statements is LIKELIEST to be false?

a. PDS is autosomal recessively inherited
b. Is often bilateral
c. Myopia is a risk factor
d. Associated glaucoma is up to 3 times more likely to occur in men as compared to women

32. A 9-year-old girl presents with progressive, painless, bilateral blurring of vision for 5 months. Her visual acuity is 6/24 in both eyes. On examination, she has yellowish sub-retinal flecks involving the posterior poles. What is the MOST likely diagnosis?

a. Fundus Flavimaculatus
b. Stargardt disease
c. Familial Drusen
d. Benign Flecked Retinal Syndrome

33. Which of the following statements is LEAST likely to be false about Best's Disease?

a. Autosomal recessively inherited disorder
b. Onset in late teens
c. Asymptomatic at onset of disease
d. Is not detectable by electrophysiology in the early stages

34. Which of the following statements regarding gyrate atrophy is MOST likely to be false?

a. Arises from mutations involving the gene which codes for ornithine aminotransferase (OAT)
b. The vast majority of patients present with peripheral visual loss in the first decade of life
c. Associated with elevated plasma and urine ornithine
d. There is robust evidence for pyridoxine as therapy to reduce the rate of progression of the disease

35. According to Royal College guidelines for the management of Hemi-retinal Vein Occlusion, the following are true EXCEPT

a. Treatment algorithm is similar to that for management of Branch Retinal Vein Occlusion
b. Grid laser should not be first line treatment for patients presenting with macular oedema within 3 months
c. A central foveal thickness of more than 250μm is an indication for treatment with dexamethasone implant (Ozurdex)
d. Re-treatment should be discontinued if the patient achieves a visual acuity of 6/12 or better

36. Which of the following statements regarding the use of Social Media by doctors is MOST likely to be true?

a. It is acceptable for a patient to contact a doctor via their personal Facebook page
b. It is acceptable to post pictures of a patient's face online
c. You must post your full name when identifying yourself as a doctor online
d. Defamatory remarks posted online are not subject to the same laws legislating written and verbal defamation

37. Regarding the following paragraph, which of the following statements is LEAST likely to be true?

According to the American College of Rheumatology, the presence of 3 of 5 of the following criteria carries a 93.5% sensitivity for giant cell arteritis (GCA).
- Age 50 years or more at disease onset
- New onset of a localized headache
- Temporal artery tenderness or decreased pulsation
- Positive temporal artery biopsy
- Elevated ESR

a. The presence of 3 or more criteria will occur in 93.5% of patients who have giant cell arteritis
b. 6.5% of patients with the disease will not have 3 or more criteria
c. The positive predictive value of testing of these criteria is likely to be higher in the general population as compared to patients presenting with suspected ischemic optic neuropathy
d. Based on the sensitivity values, these criteria are acceptable to be used as a screening test in patients presenting with suspected ischemic optic neuropathy

38. Which of the following is LEAST likely to be a differential diagnosis for arthritis in a child?

a. Kawasaki Disease
b. Juvenile Idiopathic Arthritis (JIA)
c. Septic Arthritis
d. Gout

39. Which of the following conditions is LEAST likely to cause optic disc swelling?

a. Anterior Ischemic Optic Neuropathy
b. Leber's Hereditary Optic Neuropathy
c. Demyelinating Optic Neuritis
d. Kjer Hereditary Optic Neuropathy

40. Which of the following statements regarding treatment options and prognosis for Acute Demyelinating Optic Neuritis (ADON) is MOST likely to be true?

a. Final outcome in terms of visual acuity improves after treatment with high dose intravenous methylprednisolone
b. Treatment with intravenous methylprednisolone hastens visual recovery
c. Treatment with oral prednisolone reduces the risk of recurrence as compared to no treatment
d. The presence of 1 or more white matter lesions found on magnetic resonance imaging (MRI) after an initial demyelinating event increases the 15-year risk of developing multiple sclerosis to 40%

41. Regarding the presenting features of Leber's Hereditary Optic Neuropathy (LHON), which of the following statements is MOST likely to be true?

a. Presentation is typically with bilateral, symmetrical, painless loss of vision
b. Visual acuity usually deteriorates to around 6/36 and stabilizes
c. There is optic disc swelling in the acute phase
d. Pupillary reactions are usually abnormal

42. Regarding Wernicke's encephalopathy, which of the following statements is likely to be false?

a. Is associated with vitamin B1 deficiency
b. Is characterized by a triad of ataxia, ophthalmoplegia and confusion
c. Pupil abnormalities are more common than nystagmus
d. There are often signs of peripheral neuropathy

43. Which of the following conditions is the MOST likely diagnosis for the following scenario?

A 95-year-old lady presents with a purple, painless swelling on her right upper lid. She has submandibular lymphadenopathy. Excision and eyelid reconstruction is performed and she is referred to the ENT surgeons. Histopathology of the lesion reveals small, undifferentiated cells which stain for neuron specific enolase.

a. Kaposi Sarcoma
b. Merkel Cell Carcinoma
c. Verrucous Carcinoma
d. Sebaceous Gland Carcinoma

44. A 65-year-old gentleman presents with bilateral drooping of the upper lids. On examination, he has marked atrophy of the orbital fat pads with deep upper lid sulci. Both upper lid creases are absent. Ocular motility is normal.

Lid measurements are as follows:
Palpebral apertures OD: 6 mm OS: 5 mm
Marginal Reflex Distances 1 (MRD1) OD: 1 mm OS: 0 mm
Levator function OD: 14 mm OS: 15mm

What is the MOST likely diagnosis?

a. Involutional Ptosis
b. Dermatochalasis
c. Brow Ptosis
d. Myopathic Ptosis

45. The following statements regarding Brown's syndrome are likely to be true EXCEPT?

a. Caused by a short or inelastic superior oblique tendon
b. More often acquired than congenital
c. Usually unilateral
d. There is difficulty elevating the eye in adduction

46. Regarding Choroidal Ruptures, which of the following statements is MOST likely to be false?

a. Are breaks involving the choroid, Bruch's membrane, and retinal pigment epithelium (RPE)
b. Higher chance of occurring in contused eyes as compared to eyes that have suffered an open globe injury
c. Indirect ruptures are usually perpendicular to the limbus
d. Visual prognosis is generally good unless the macula is involved

47. Regarding lattice degeneration, which statement is LEAST likely to be true?

a. Seen in 8% of patients with rhegmatogenous retinal detachments
b. May be seen with increased frequency in patients with Marfan's syndrome.
c. Vitreous is strongly adherent to the lesion edges
d. Should be treated in a pseudophakic fellow eye in patients with a rhegmatogenous retinal detachment.

48. Select the MOST appropriate treatment option for the following scenario.

A 67-year-old gentleman with advanced primary open angle glaucoma presents with sudden onset blurring of vision in his left eye for 2 days. On examination, his best-corrected visual acuity is OD 6/12 OS 6/48. There is no relative afferent pupillary defect. There is early cataract in both eyes. IOP is 19 in both eyes on 3 anti glaucoma drops.
The fundus of the right eye is normal, cup disc ratio 0.8. The fundus of the left eye shows widespread intraretinal haemorrhages and dilated veins in all 4 quadrants, cup disc ratio 0.9.
There is macular oedema with foveal thickness of 420μm.

a. Intravitreal ranibizumab
b. Intravitreal dexamethasone implant
c. Intravitreal triamcinolone
d. Macular grid laser

49. The following are surgical procedures to improve abduction in a patient with 6th nerve palsy EXCEPT?

a. Botulinum toxin injection to the ipsilateral medial rectus
b. Medial rectus recession
c. Jensen Transposition
d. Knapp Transposition

50. For the following clinical scenario, please choose the MOST likely treatment option

A 25-year-old lady with myopia presents with acute onset flashes and floaters in her right eye associated with an inferior visual field defect.
On examination, her visual acuity is 6/6. There is a bullous supero-temporal rhegmatogenous retinal detachment with 2 breaks at 10 and 11 o clock, visible only with indentation. There is no proliferative vitreoretinopathy (PVR). There is no Weiss ring.
She is very anxious, and is keen to avoid repeat surgery as much as possible.

a. Scleral buckling and external drainage
b. Pneumatic retinopexy
c. Pars plana vitrectomy and silicone oil tamponade
d. Pars plana vitrectomy and SF6 gas tamponade

51. A 65-year-old lady presents with epiphora. On examination she has right lower lid ectropion. There is lateral canthal tendon laxity. What is the MOST appropriate surgical option?

a. Wedge Excision
b. Inverting sutures
c. Tarsal Strip
d. Blepharoplasty

52. According to Royal College of Physicians joint recommendations on osteoporosis prophylaxis for patients on long term corticosteroid therapy, which of the following statements is false?

a. All patients above 65 years should be treated
b. All patients below 65 years with fragility fractures should be treated
c. Patients below 65 years with a T score of <-1.5 standard deviations (SD) should be treated
d. Patients below 65 years with a T score between 0 and -1.5 SD should be treated.

53. Regarding parental consent for a child undergoing a surgical procedure, which of the following statements is INCORRECT?

a. The child's genetic mother is able to give consent
b. Grandparents are not able to give consent if a child's parents are unavailable
c. The child's genetic father who was married to the mother at the birth but was divorced at the time of consultation and whose name is not on the birth certificate is able to give consent
d. The child's unmarried genetic father who was present at the birth in 2005 and whose name is on the birth certificate is able to give consent

Questions 54 and 55 concern the table below.

The following table reveals data from a randomized controlled trial designed to determine the effect of an experimental intravitreal therapy vs. standard therapy on age related macular degeneration.

	Final Visual Acuity <6/60	Final Visual Acuity >6/60	Total
Standard Therapy (X)	12	48	60
Experimental Therapy (Y)	15	90	105

54. What is the absolute risk reduction of eventually developing a visual acuity of <6/60?

a. 0.2
b. 0.14
c. 0.06
d. 0.6

55. What is the number needed to treat (NNT) from the table above?

a. 17
b. 1.7
c. 0.17
d. 0.017

56. Which of the following biometric formulas is recommended for calculating the power of an intraocular lens for an eye with an axial length of 21.72 mm?

a. SRK/T
b. Holladay 1
c. Holladay 2
d. Hoffer Q

57. Regarding viscoelastics, which statement is MOST likely to be true?

a. Sodium hyaluronate (SH) is derived from sharks fins
b. Hydroxypropylmethylcellulose (HPMC) is extracted from wood pulp.
c. Chondroitin sulphate (CS) has a higher molecular weight as compared to sodium hyaluronate
d. HPMC is an effective intraoperative tool for anterior chamber maintenance during capsulorhexis

58. Which of the following is NOT a risk factor for expulsive haemorrhage during cataract surgery?

a. Hypertension
b. Extra-capsular cataract extraction
c. Vitreous loss
d. Intraoperative bradycardia

59. The following are features of optic disc pits EXCEPT?

a. Usually bilateral
b. Fluid from the pit may cause a macular retinoschisis
c. Serous macular detachment is a sight threatening complication
d. Fluorescein angiography typically reveals pooling beneath a serous detachment

60. Regarding ocular manifestations of herpes simplex virus (HSV), which of the following statements is MOST likely to be true?

a. Ocular infection is caused primarily by HSV 2
b. There is robust evidence for stress as a trigger of recurrent eye disease
c. Primary herpes infection of the eye manifests predominantly as a unilateral follicular blepharoconjunctivitis
d. Epithelial keratitis often progresses to cause stromal involvement if not treated promptly with topical anti-viral agents

61. The following statements regarding primary optic nerve sheath meningioma (PONSM) are false except?

a. Exhibit a male preponderance
b. Exhibits accelerated growth in pregnancy
c. Are more aggressive in older patients
d. MRI scans are unable to reliably distinguish between PONSM and optic nerve glioma

62. A 60-year-old lady with underlying diabetes and hypertension and history of previous stroke presents to eye casualty because she has been bumping into objects on her right side with increasing frequency. On examination, she has visual acuity of OD: 6/6 OS: 6/6 and normal colour vision. Her optic discs are normal.
Systemic examination reveals no cerebellar signs, no limb weakness and no peripheral loss of sensation.

Where is the offending lesion likely to be?

a. Optic nerve
b. Thalamus
c. Internal Capsule
d. Occipital lobe

63. The following are risk factors for developing mucormycosis EXCEPT?

a. Uncontrolled Diabetes
b. Chronic Sinusitis
c. Intravenous Drug Abuse
d. Renal Transplantation

64. Regarding congenital nasolacrimal duct (NLD) obstruction, which of the following statements is MOST likely to be false?

a. Obstruction is often proximal to the nasolacrimal sac
b. 90% of cases resolve spontaneously by 1 year
c. Probing carries a 90% success rate
d. Dacryocystorhinostomy (DCR) is indicated should probing fail to treat the epiphora

65. Which of the following statements regarding the presentation of Duane's Syndrome is INCORRECT?

a. Isolated in 30% of cases
b. Associated with Morning Glory Syndrome
c. Typically presents with esotropia
d. Less commonly involves the right eye

66. Which of the following statements regarding rod monochromatism is TRUE?

a. Is X-linked
b. Presents with nystagmus from ages 5 years onwards
c. Macula often shows pigmentary changes
d. There is usually a central scotoma with normal peripheral fields

67. Which of the following statements regarding the clinical differences between neurogenic and restrictive strabismus is MOST likely to be false?

a. Neurogenic strabismus demonstrate monocular ductions that exceed versions whilst restrictive strabismus tend to demonstrate ductions that are similar to versions
b. Intraocular pressure remains the same in neurogenic strabismus but will increase in the direction of limitation in restrictive strabismus
c. Inner and outer fields of the Hess chart are equally compressed in restrictive strabismus, while the outer fields are compressed to a greater degree compared to the inner fields in neurogenic strabismus
d. Forced duction test is negative in acute neurogenic strabismus but positive in restrictive strabismus

68. Silicone oil insertion is often necessary as an adjunct during surgery to repair retinal detachments caused by acute retinal necrosis (ARN) for the following reasons EXCEPT?

a. It is often impossible to identify the primary break
b. Post-operative control of haemorrhage is better than with gas
c. The retina is often thinned, increasing the risk of new break formation
d. Proliferative vitreoretinopathy (PVR) is often present

69. Which of the following statements regarding posterior capsular opacification (PCO) is MOST likely to be INCORRECT?

a. Young patients are at increased risk
b. Non - angulated posterior chamber intraocular lenses (IOLs) may increase predisposition towards PCO formation
c. Round edged optics are superior to truncated edged optics in inhibiting posterior migration of lens epithelial cells (LEC)
d. Hydrophobic acrylic material is superior to hydrophilic material in promoting adhesion of the optic to the capsular bag and minimising the presence of a migration plane for LES

70. Where is the likeliest site of the primary break in the following scenario?

A 50-year-old gentleman with myopia presents with an inferior retinal detachment in his right eye. The detachment extends from 5 to 10 o clock and involves the macula.

a. 6 o clock
b. 9 o clock
c. Above the horizontal meridian
d. Near 12 o clock

71. Regarding band keratopathy, which of the following statements is MOST likely to be true?

a. Caused by calcium deposits within Descemet's membrane
b. Stain red with Von Kossa staining
c. Symptomatic in the early stages
d. Starts at 3 and 9 o clock to progress centrally

72. Which of the following White Dot Syndromes typically manifests unilaterally?

a. Acute Posterior Multifocal Placoid Pigment Epitheliopathy (APMPPE)
b. Multiple Evanescent White Dot Syndrome (MEWDS)
c. Punctate Inner Choroidopathy
d. Birdshot Choroidopathy

73. The following glaucomatous disorders may be treated by a patent laser iridotomy EXCEPT?

a. Chronic uveitis with extensive posterior synechiae
b. Iridoschisis
c. Iridocorneal Endothelial Syndrome (ICE)
d. Silicone Oil in an aphakic eye

74. Which of the following corneal dystrophies does not commonly demonstrate post-operative recurrence?

a. Lattice Dystrophy
b. Macular Dystrophy
c. Granular Dystrophy
d. Avellino Dystrophy

75. Regarding the prognosis of a patient with conjunctival melanoma, which of the following statements is MOST likely to be false?

a. The overall mortality rate is 25%
b. Tumours involving the caruncle carry an unfavourable prognosis
c. Histopathologic examination revealing spindle cells carries a worse prognosis compared to mixed cell
d. Histologic evidence of lymphatic spread carries an unfavourable prognosis

76. The following statements regarding investigations for suspected direct carotid cavernous fistulae are true EXCEPT?

a. The definitive investigation is computed tomographic (CT) angiography
b. Extraocular muscles may be enlarged on CT scans
c. Dilation of both superior ophthalmic veins are seen
d. The cavernous sinus is enlarged on contrasted CT scans

77. Which of the following organisms is MOST likely to be the offending pathogen in the following scenario?

A 65-year-old Scottish farmer, while clearing brush, is poked in the eye by a tree branch. He subsequently develops eye redness and discomfort. On examination, he has a central feathery corneal opacity with a dry, elevated appearance.

a. Candida spp.
b. Fusarium spp.
c. Pseudomonas aeruginosa
d. Aspergillus spp.

78. The following statements regarding albinism are true EXCEPT?

a. Pure ocular albinism is an x-linked disorder
b. Features of ocular albinism include photophobia and nystagmus
c. Foveal hypoplasia is a typical finding
d. Ocular albinism is more common than oculocutaneous albinism

79. Which of the following statements regarding Coat's Disease is INCORRECT?

a. There is no racial predilection
b. Unilateral in 80%
c. Affects females more than males
d. May be confused with retinoblastoma

80. Which of the following corneal dystrophies does not typically present with recurrent corneal erosions?

a. Granular Dystrophy
b. Type 3 Lattice Dystrophy
c. Reis Buckler Dystrophy (RB)
d. Epithelial Basement Membrane Dystrophy (EBMD)

81. According to the findings of the following investigations, what is the MOST likely diagnosis?

A 60-year-old lady presents with an elevated orange choroidal lesion with high internal reflectivity on ultrasound.
There is absence of intralesional blood vessels on fluorescein angiography with early hypofluorescence and late pinpoint hyperfluorescence.

a. Choroidal Melanoma
b. Choroidal Metastasis
c. Choroidal Haemangioma
d. Melanocytoma

82. The following features can be used to differentiate a unilateral from bilateral 4th nerve palsy EXCEPT?

a. Chin down head posture
b. Double Maddox rod test showing > 20 degrees of excyclotorsion
c. Reversing hyperdeviation on lateral gaze
d. Prominent V pattern

83. Which of the following is LEAST likely to be a feature of Pellucid Marginal Degeneration (PMD)?

a. Progressive irregular astigmatism
b. Painless
c. Central corneal steepening
d. Inferior corneal thinning

84. The following genetic markers are associated with a worse prognosis for uveal melanoma EXCEPT?

a. Chromosome 8q Gain
b. Monosomy 3
c. Chromosome 6p Gain
d. Mutations involving the gene for BRCA associated Protein 1 (BAP1)

85. A relative afferent pupil defect (RAPD) is MOST likely to be seen with which of the following lesions?

a. Stroke involving the thalamus
b. Glioma of the temporal lobe
c. Pituitary microadenoma
d. Giant anterior cerebral artery aneurysm

86. Which of the following conditions is MOST likely to present with reduced visual acuity in childhood?

a. Fundus albipunctatus
b. Stargardt's Disease
c. Oguchi Disease
d. Bardet Biedl Syndrome

87. Regarding Myasthenia Gravis (MG), which of the following statements is MOST likely to be false?

a. Characterised by the presence of antibodies towards post synaptic acetylcholine receptors (AchR)
b. Ocular MG is the presenting feature in 50% of patients with MG.
c. The majority of patients presenting with ocular MG remain with purely ocular weakness
d. The majority of patients with generalized MG eventually have ocular involvement at some point in their illness

88. Which of the following statements pertaining to Reiter's Syndrome is MOST likely to be correct?

a. Acute anterior uveitis occurs in 50% of patients with Reiter's
b. Reiter's may be post infectious, following a bout of gonococcal urethritis
c. Affects males more than females
d. Aortic incompetence is a frequent, life threatening complication

89. Regarding case control studies, which of the following statements is FALSE?

a. Samples are chosen based on the presence or absence of risk factors for disease
b. Observational study design
c. Often retrospective
d. Allows establishment of cause-and-effect

90. According to the joint Royal College of Ophthalmologists (RCOphth) and Royal College of Paediatrics and Child Health (RCPCCH) Guidelines, when should premature babies born before 27 weeks undergo their initial screen for retinopathy of prematurity (ROP)?

a. 29-30 weeks post-menstrual age
b. 30-31 weeks post-menstrual age
c. 31-32 weeks post-menstrual age
d. 4-5 weeks post-natal age

Paper 3 Answers and Discussion

1. Answer: d

Discussion: Warkany Syndrome 2 or Trisomy 8 is a very rare, ophthalmologically significant syndrome. Ocular features include hypertelorism, deep-set eyes, and bilateral corneal opacities.

Source:
1. Agrawal A, Agrawal R. Warkany Syndrome: A Rare Case Report. *Case Reports in Pediatrics*. Volume 2011 (2011), Article ID 437101.

2. Answer: d

Discussion: CJD transmission via instrumentation and/or tissue transplantation in anterior segment surgery carries a lower risk as compared to the other factors.
Thus, as per Royal College recommendations, there is no need for any deviation in practice whilst performing anterior segment surgery in a patient deemed to be at risk for prior infection with CJD

Source:
1. Royal College of Ophthalmologists Guidance on CJD in conjunction with the Advisory Committee on Dangerous Pathogens TSE Working Group 2010

3. Answer: b

Discussion: Development of myopia is multifactorial and tends to develop between the ages of 6 and 14. Risk factors include prematurity, low birth weight, a positive family history, scholastic achievement and constant near work.

Source:
1. Saw SM, Katz J, Schein OD, Chew SJ, Chan TK. Epidemiology of Myopia. *Epidemiologic Reviews*. Vol 18, No 2. 1996

4. Answer: c

5. Answer: b

Discussion: Retinoblastoma arises from mutations to the Rb1 gene on the long arm of Chromosome 13 in immature retinal cells.

Two transmission patterns develop – *Hereditary retinoblastoma*, where an increased risk of developing retinoblastoma is transmitted to the affected individual's children; and *non-hereditary retinoblastoma*, where there is no transmission of increased risk.

Hereditary retinoblastoma occurs when all the cells of an affected child carry an abnormal copy of the Rb1 gene. Up to 90% of these children acquire a germ line-mutation during the embryonic stage. The affected child then carries the mutation in all their body cells, which then makes it relatively easier to develop retinoblastoma when a developing retinoblast acquires a second mutation
The remaining 10% of these children acquire the mutation from a parent, thus often demonstrating a positive family history of retinoblastoma
Hereditary retinoblastoma accounts for 40% of all cases, and tends to be bilateral in up to 90%.

Non-hereditary retinoblastoma occurs when both the Rb1 genes undergo mutation in a single, developing retinoblast. There is no increased risk (above background risk) of developing retinoblastoma in the affected child's children as they do not carry the faulty gene in their body's cells.
Non-hereditary retinoblastoma accounts for 60% of all cases and tend to be unilateral.

6. Answer: d

Discussion: Pseudomonas is an obligate aerobic bacteria, whereas **C**lostridia, **B**acteroides and **A**ctinomyces are obligate anaerobic, gram positive rods. (mnemonic: **C**annot **B**reathe **A**ir). Anaerobes are infrequent causes of ocular infection including conjunctivitis, canaliculitis, and post-operative endophthalmitis.

Source:
1. Aggarwal R, Chaudhry R, Mathur SP, Talwar V. Bacteriology of ophthalmic infections with special reference to anaerobes. *Indian J Med Res* 1992 May;95:148-51.

7. Answer: c

Discussion: Expected complications to be expected during phacoemulsification for mature, brunescent cataracts include poor capsular visualization, high intra-lenticular pressure which increases the risk of radial tears, weak zonules, and floppy posterior capsules, which increases the risk of capsular rupture in the setting of post occlusion surges.

8. Answer: b

Discussion: Fleischer lines are seen in keratoconus, Ferry's lines in patients with trabeculectomies, and Hudson Stahli lines are seen in the aging cornea. Iron lines are typically innocuous and may be observed with cobalt blue light.

9. Answer: a

Discussion: Glaucoma in ICE syndrome occurs in approximately 50% of patients and tends to be more severe in patients with predominantly Cogan Reese or iris atrophy. Glaucoma is due to high peripheral anterior synechiae extending to Schwalbe's line, causing secondary angle closure. Augmented trabeculectomies are often first line options for treatment, with success rates reported in the region of 60-70% at 1 year.

Source:
1. Laganowski HC, Kerr Muir MG, Hitchings RA. Glaucoma and the iridocorneal endothelial syndrome. *Arch Ophthalmol.* 1992;110:346-350.
2. Doe EA, Budenz DL, Gedde SJ, et al. Long-term surgical outcomes of patients with glaucoma secondary to the iridocorneal endothelial syndrome. *Ophthalmology.* 2001;108:1789-1795.

10. Answer: c

Discussion: The risk of developing retinal neovascularization in an ischaemic BRVO with more than 5 disc diameters of capillary fall out is 36% within the first 6-12 months

Source:
1. Branch Vein Occlusion Study Group. Branch Vein Occlusion Study. *Am J Ophthalmol.* 1984 Sep 15;98(3):271-82.

11. Answer: c

Discussion: The commonest field defects in ethambutol toxicity include central or centrocaecal scotomas. Other defects such as bitemporal defects and constriction of the visual field have also been reported.

Source:
1. Lim SA. Ethambutol associated optic neuropathy. *Ann Acad Med Singapire.* Apr 2006;35(4):274-8

12. Answer: c

Discussion: Cutaneous melanoma is associated with ultraviolet (UV) exposure, cutaneous disorders such as xeroderma pigmentosum, immunosuppession, male sex, older age, fair skin, and positive family history. The incidence of cutaneous melanoma in Australia has been reported to be 10-20 times that of Europe – underscoring the significance of decreasing latitude and subsequent UV exposure. The use of tanning beds has also been implicated. 4 main subtypes exist, of which the most common is the superficial spreading melanoma. The subtypes have no bearing on overall prognosis.

Source:
1. www.who.int/uv/health/uv_health2/en/index1.html

13. Answer: b

Discussion: MPS Types 1(Hurler), 4(**M**orquio), 6 (**M**aroteaux Lamy) and 7 (**S**ly) present with cloudy corneas. Mnemonic: **Cloudy** skies **M**ake **M**any **H**url and **S**ulk.

Source:
1. Denniston AKO, Murray PI. 2006. Oxford Handbook of Ophthalmology. Oxford. Oxford University Press.

14. Answer: b

Discussion: The above scenario first of all demonstrates *hypermetropia* and an *esotropia* that manifests only with glasses off. This points to a *refractive esotropia* that appears to be fully accommodative as glasses completely correct the deviation.

In contrast, Partially Accommodative Esotropias often manifest a degree of convergence excess and are not fully corrected with glasses.
Convergence Excess Esotropia often presents with minimal hypermetropia (in contrast to the significant hypermetropia in this scenario) – but an increased AC:A ratio necessitates bifocal segments.

Source:
1. MacEwen C, Gregson R. 2003. *Manual of Strabismus Surgery*. London. Elsevier Limited.

15. Answer: c

Discussion: Risk factors for traumatic endophthalmitis in open globe injuries include delayed surgical closure > 24 hours, soil contamination, retained intraocular foreign bodies, lens injury, and the presence of filtering blebs.
Shrapnel injuries are not risk factors as the fragments undergo heat sterilisation.

Source:
1. Essex RW, Yi Q, Charles PG, Allen PJ (2004) Post-traumatic endophthalmitis. *Ophthalmology* 111: 2015–2022
2. Coyler M, Weber E, Weichel E, Dick J, Bower K, Ward T, Haller J (in press) Delayed intraocular foreign body removal without endophthalmitis during Operations Iraqi and Enduring Freedom. Ophthalmology.
3. Thompson JT, Parver LM, Enger CL, Mieler WF, Liggett PE (1993) Infectious endophthalmitis after penetrating injuries with retained intraocular foreign bodies. *Ophthalmology* 100: 1468–1474

16. Answer: b

Discussion: FEVR is a hereditary vitreoretinopathy which is predominantly autosomal dominant. It is caused by poorly perfused, hyper-permeable peripheral retinal vessels that leak and bleed, causing vitreous contraction, exudative and tractional detachment.
Refraction often reveals emmetropia, and thus, is of little use in allowing differentiation from ROP. Massive exudation may cause confusion with Coats, which is however, unilateral. Furthermore, the exudation in FEVR is often combined with a tractional component.

Source:
1. Ryan SJ, Schchat AP, Hinton DR, Wilkinson CP. 2005. *Retina*. 4th Ed. Elsevier.

17. Answer: b

Discussion: Specular microscopy is the most accurate way of examining the endothelium. It is non-invasive, capturing microphotographs by way of light reflection from the endothelial-aqeous interface. Abnormalities such as reduced cell counts, polymegethism or pleiomorphism may be signs of pathology or more commonly, age.

18. Answer: d

Discussion: Shellfish allergies are unrelated to allergies to contrast medium

Source:
1. American College of Radiology - www.acr.org

19. Answer: d

Discussion: Various methods of disinfection have been described for reusable tonometer tips.
A study published in *Eye* in 2007 showed that, of the 4 options described above, a 3% hydrogen peroxide soak for 3 minutes was most effective in eliminating the growth of both bacteria and fungi.

Source:
1. Cilino S. et al. Tonometers and infectious risk: myth or reality? Efficacy of different disinfection regimens on tonometer tips. *Eye* (2007) 21, 541–546

20. Answer: d

Discussion: The Ishihara Pseudoisochromatic Plates utilize numbers constructed from dots of different hue to detect red-green colour defects. The dots are of the same contrast and under ideal lighting conditions appear to have similar luminance, thus ensuring that only the hue differs between dots.
Ideal lighting conditions are either a daylight filled room, or electrical lighting closely resembling daylight.
The lower limit of acuity in to read the test plate is 6/60. Test results are scored out of 14. Patients with history of acute demyelinating optic neuritis with normal or near normal acuity may have a markedly abnormal test.
Prolonged exposure of the test plates to direct sunlight may cause fading of the colours and loss of test accuracy.

Source:
1. Clarke C, Howard R, Rossor H, Shorvon SD. (2011) *Neurology*. John Wiley and Sons.

21. Answer: c

Discussion: Screening for Graves disease is easiest done with thyroid hormone assays. Serologic studies support the diagnosis of Graves disease but may be positive in the absence of any clinical symptoms or signs. TSH antibodies are seen in 40-95% of patients with TED. Thyroid peroxidase and thyroglobulin antibodies are seen less frequently. Antibodies to collagen XIII are highly indicative of active TED. Haematologic findings include normocytic anaemia and a relative lymphocytosis.

Source:
1. De Bellis A, Sansone D, Coronella C, et al. Serum antibodies to collagen XIII: a further good marker of active Graves' Ophthalmopathy. *Clin Endocrinol (Oxf)*. Jan 2005;62(1):24-9.
2. Yeung S-CJ. *Graves Disease*. emedicine.medscape.com

22. Answer: b

Discussion: Differentials for subretinal solid lesions include naevi, melanomas, melanocytomas, metastases, osteomas, haemangiomas and retinal pigment epithelial (RPE) lesions such as Congenital Hypertrophy of the RPE (CHRPE). Often, ultrasonography, FA, indocyanine angiography (ICG) and to a lesser extent, MRI, will aid in diagnosis.

The ultrasound and angiographic findings strongly suggest a choroidal haemangioma. In contrast, while melanomas may present with macular oedema, ultrasonography typically shows internal hyporeflectivity and early hypofluorescence. Angiographic findings in melanomas may also reveal a dual circulation.
A choroidal metastatic lesion would present with internal hyper-reflectivity on ultrasonography and early hypofluorescence/lack of internal vasculature on angiogram.

23. Answer: c

Discussion: The Titmus Fly test is a measure of gross stereopsis of up to 3000 seconds of arc

24. Answer: d

Discussion: Movement of drug from the tear reservoir into the eye is mainly by passive diffusion, driven by the drug concentration gradient between the tear cul-de-sac and the intraocular spaces.
The significant barriers to this movement comprise the tight junctions between corneal epithelial cells, the water rich corneal stroma, and tight junctions between corneal endothelial cells.
The perilimbal conjunctiva is permeable to small, water soluble molecules, and thus offers an alternative trans scleral route for diffusion.
Contact duration with the cornea is critical for drug penetration; this is aided by increased viscosity of the drug vehicle, increased drug concentration, avoidance of blinking and punctal occlusion.
Drug factors that aid penetration include degree of ionization (non-ionized molecules cross the epithelium and endothelium more readily – the converse is true for the corneal stroma) and smaller molecule size.

25. Answer: b

Discussion: Cyclophosphamide is an alkylating agent. Methotrexate and Mycophenolate are both antimetabolites that impair DNA synthesis.
Calcineurin is a intracellular protein that initiates production of IL-2 by activation of nuclear transcription factors. It is the target of the group of drugs called transcription factor inhibitors that include ciclosporin and tacrolimus which are often used as anti rejection drugs. Ciclosporin causes nephrotoxicity in up to a third of patients.

Source:
1. *emedicine.medscape.com*

26. Answer: d

Discussion: Apraclonidine is a direct acting alpha-2 agonist. It lowers intraocular pressure by a combination of reduced aqeous production, improved trabecular outflow and decreased episcleral venous pressure. It is effective in the short term. Topical sensitivity and tachyphylaxis limits long term use. Brimonidine has a superior side effect profile.

27. Answer: b

Discussion: IFIS is associated strongly with alpha-1-antagonists primarily used for treatment of Benign Prostatic Hypertrophy, in particular, Tamsulosin. The drugs reduce smooth muscle tone of the iris, causing a classic triad of iris stromal billowing, prolapse through phaco and side port incisions, and *progressive, intraoperative miosis.*
An authoritative White Paper by the ASCRS/ESCRS in 2008 concluded that there was no robust evidence that stopping Tamsulosin effectively reduced the risk of IFIS.

Source:
1. Chang et al. ASCRS White Paper: Clinical review of intraoperative floppy-iris syndrome. *J Cataract Refract Surg* 2008; 34:2153–2162 Q 2008 ASCRS and ESCRS

28. Answer: d

Discussion: Adult inclusion conjunctivitis is caused by serotypes D-K, is sexually transmitted, and may present with non-tender lymphadenopathy. Swabs should be taken for cultures and patients referred to a genitourinary clinic for systemic examination and treatment. First line treatment is typically topical chloramphenicol and oral doxycycline. For women in reproductive age group, oral erythromycin should be considered first line systemic therapy.

29. Answer: d

Discussion: SJS tends to affect younger patients, and has been reported to occur with a mean age of 25-47 depending on the series. OCP tends to affect patients in their sixties, with a female: male ratio of 2:1.
Cutaneous involvement is the norm in SJS, with an involved body surface area (BSA) of up to 10%. Cutaneous involvement is rare in OCP.
Systemic immunosuppression in OCP is the mainstay of therapy, while still controversial in SJS.

Source:
1. Power WJ, Ghoraishi M, Merayo-Lloves J, Neves RA, Foster CS
Analysis of the acute ophthalmic manifestations of the erythema multiforme/Stevens-Johnson syndrome/toxic epidermal necrolysis disease spectrum.Ophthalmology. 1995;102(11):1669.
2. Patterson R, Miller M, Kaplan M, Doan T, Brown J, Detjen P, et al. Effectiveness of early therapy with corticosteroids in Stevens-Johnson syndrome: experience with 41 cases and a hypothesis regarding pathogenesis. Ann Allergy. 1994;73(1):27.

30. Answer: c

Discussion: The causes of deposition keratopathies may be divided into *infectious* and *non-infectious* causes.
Non-infectious causes include: Wilson's Disease, Argyria, Chrysiasis, and protein deposits e.g immunoglobulin deposits in Waldenstrom's or lymphomas.
Infectious causes include biofilms caused by *Streptococcus viridans*, and less commonly *Staphylococci spp.*
Vortex Keratopathy is often innocuous, and may be associated with the mnemonic CATFISH
(**C**hlorpromazine, **A**miodarone, **T**amoxifen, **F**abry's Disease, **I**ndomethacin, **S**uramin and **H**ydroxychloroquine)

Source:
1. Sharma N, Vajpayee RB, Pushker N, Vajpayee M. Infectious crystalline keratopathy. *CLAO J.* 2000 Jan;26(1):40-3

31. Answer: a

Discussion: PDS is autosomal dominantly inherited, affects men as well as women, but is 3 times more likely to cause glaucoma in men as compared to women. Risk factors include myopia, which is seen in 80% of eyes presenting with PDS. Features include deep anterior chambers, pigment on the endothelium, and characteristic spoke like iris transillumination defects.

Source:
1. www.glaucoma.org

32. Answer: b

Discussion: Stargardt disease is the most likely diagnosis, given the age and rapid deterioration of vision. It is the most commonly inherited macular dystrophy, presenting in childhood with rapid deterioration of vision and bilateral orange posterior pole flecks. Fundus Flavimaculatus tends to present in older patients with relative preservation of vision. Famililal drusen may be a differential diagnosis, however, symptoms are often mild.

33. Answer: c

Discussion: Best's is a rare, autosomally dominant disorder characterized by a mutation in the gene coding for the protein *bestrophin,* leading to lipofuscin collecting between the RPE and retina at the posterior pole. The disease usually has its onset between the ages of 3-15, progresses through 5 well defined stages, is asymptomatic in the early stages but may be detected via EOG in the pre-vitelliform stage.

Source:
1. Royal National Institute for the Blind, www.rnib.org.uk

34. Answer: d

Discussion: Gyrate atrophy is an autosomal recessively inherited choroidal dystrophy stemming from mutations in the OAT gene which leads to excessive levels of plasma and urine ornithine. Patients tend to present in the first decade with nyctalopia and peripheral visual loss. Fundus findings include well demarcated areas of choroidal atrophy which start peripherally and then proceed centrally with relative sparing of the macula, disc and vessels. In about 5% of patients, large doses of vitamin B6 have been proven to lower levels of ornithine in the blood without any significant effect on the rates of disease progression

Source:
1. Kaiser-Kupfer MI, Caruso RC, Valle D. Gyrate Atrophy of the Choroid and Retina Further Experience With Long-term Reduction of Ornithine Levels in Children. *Arch Ophthalmol.* 2002;120(2):146-153

35. Answer: d

Discussion: The treatment algorithm for hemiretinal vein occlusion is similar to that for branch retinal vein occlusion. The management largely depends on the timing of presentation and the state of the macula. Patients presenting with macular oedema with a central foveal thickness of more than 250µm within 3 months should be treated with either intravitreal dexamethasone implant (Ozurdex) or intravitreal ranibizumab. Grid laser can be considered for patients presenting with macular oedema after 3 months. Criteria for discontinuation of treatment should be considered once the patient's visual acuity improves to better than 6/7.5, central foveal thickness of less than 250µm or the patient develops hypersensitivity.

Source:
1. Royal College of Ophthalmologists Interim Guidelines on Retinal Vein Occlusion 2010

36. Answer: c

Discussion: When identifying yourself as a doctor online it is essential to post your full name as it is reasonable for non-medical readers to regard your remarks as being representative of the opinions of the medical profession as a whole. Defamatory remarks posted online are subject to the same laws legislating written and verbal defamation

Source:
1. General Medical Council Guidelines regarding *Doctors Use of Social Media* 2013

37. Answer: c

Discussion: A sensitivity of 93.5% in the presence of 3 or more criteria suggests that 93.5% of patients who have GCA will fulfill 3 or more criteria. Conversely, the false negative rate, being 1-sensitivity, suggests that there will be a 6.5% false negative rate, i.e 6.5% of patients who have the disease will NOT have 3 or more criteria.

Positive predictive value (PPV) is the probability of having a condition, given a positive test. PPV is dependent upon the prevalence of disease in the population surveyed. Therefore, the PPV of 3 or more criteria in identifying patients with GCA is likely to be higher in patients presenting with suspected ischemic optic neuropathy as compared to the general population. A sensitivity of >90% and the fact that they are largely non-invasive, makes them excellent screening criteria.

Source:
1. American College of Rheumatology Guidelines 1990 for diagnosis of GCA

38. Answer: d

Discussion: Gout is uncommon in children. Differentials for an arthritis presenting in childhood would include JIA, Kawasaki Disease, Septic Arthritis, and haematologic malignancies.

39. Answer: d

Discussion: Disc swelling is classical in anterior ischemic optic neuropathy, and may be a presenting feature in 1/3 of patients presenting with acute demyelinating optic neuritis.
Early swelling of the nerve fiber layer is a typical feature of Leber's, while Kjer tends to present with optic disc pallor, often temporal.

40. Answer: b

Discussion: While the final outcome in terms of visual acuity does not differ regardless of treatment, intravenous methylprednisolone hastens visual recovery by a few weeks. Oral prednisolone is contraindicated as it confers no benefit and is associated with an increased recurrence rate as compared to no treatment. As per 15 year results from the Optic Neuritis Treatment Trial, the 15-year risk of developing multiple sclerosis after an initial demyelinating event increases from 25% to 72% if one or more white matter lesions are seen on MRI as compared to no lesions.

Source:
1. The Optic Neuritis Study Group. Multiple Sclerosis Risk After Optic Neuritis: Final Optic Neuritis Treatment Trial Follow-up. *Arch Neurol.* 2008;65(6):727-732.

41. Answer: c

Discussion: Presentation of LHON is typically with painless, sequential, sudden blurring of vision in healthy young adult males. Acutely, there is optic disc swelling with subsequent late pallor. Pupil reactions are usually normal. Visual acuity usually stabilizes at less than 6/60.

42. Answer: c

Discussion: Wernicke's encephalopathy is characterized by a triad of ataxia, ophthalmoplegia and confusion caused by chronic thiamine (vitamin B1) deficiency. Thiamine is critical for cerebral carbohydrate metabolism, and chronic deficiency may lead to neuronal death. The commonest ocular features are nystagmus, bilateral lateral rectus palsies and conjugate gaze palsies. Less commonly, ptosis and anisocoria may be seen. There are signs of peripheral neuropathy in around 80% of patients, with weakness, foot drop or loss of proprioception.

Source:
1. Donnino MW, Vega J, Miller J, et al. Myths and misconceptions of Wernicke's encephalopathy: what every emergency physician should know. *Ann Emerg Med.* Dec 2007;50(6):715-21.

43. Answer: b

Discussion: Merkel cell carcinomas are rare, cutaneous malignancies which typically affect the periorbital region and are frequently under-diagnosed due to lack of clinical symptoms. They arise from the Merkel cell, which has a yet to be determined function and has both epidermal and neuroendocrine features, leading to hypotheses that it is a precursor cell capable of differentiating along either lineage. They present as purple, non-tender nodules that metastasize primarily via the lymphatics. Diagnosis is difficult, as the tumors are frequently undifferentiated and thus indistinguishable from other undifferentiated tumors. Immunohistochemical staining is often required, the hallmark marker being neuron specific enolase (NSE).

Source:
1. Gu J, et al. Immunostaining of neuron-specific enolase as a diagnostic tool for Merkel cell tumors. *Cancer* 1983 Sep 15;52(6):1039-43.

44. Answer: a

Discussion: The features suggest a diagnosis of involutional ptosis, given the age, the severity of the ptosis, as well as preserved levator function. Dermatochalasis and Brow Ptosis are causes of pseudo-ptosis that would present with normal lid measurements.

45. Answer: b

Discussion: Brown's Syndrome is rarely acquired.

Source:
1. Parks MM, Brown M. Superior oblique tendon sheath syndrome of Brown. *Am J Ophthalmol*. Jan 1975;79(1):82-6.

46. Answer: c

Discussion: Choroidal ruptures are breaks involving the choroid and extending through Bruch's membrane and the RPE. They may be *direct,* occurring at the site of impact, or *indirect,* occurring often at the posterior pole of the eye where shockwaves transmitted through the walls of the eye coalesce. They are seen more often in contused (8%) as compared to ruptured eyes (1%). Indirect ruptures are often concentric to the optic disc. Visual prognosis is usually good unless the ruptures involve the fovea or if they are complicated with secondary choroidal neovascularisation.

Source:
1. Ferenc Kuhn. 2008. *Ocular Traumatology*. Springer.
2. Raman SV, Desai UR, Anderson S, Samuel MA (2004) Visual prognosis in patients with traumatic choroidal rupture. *Can J Ophthalmol* 39: 260–266

47. Answer: a

Discussion: Lattice degeneration may be seen in up to 30% of patients with rhegmatogenous retinal detachments. It occurs with increasing frequency in patients with connective tissue disease e.g Marfan's or Stickler. Acute posterior vitreous detachments tend to cause horseshoe tears extending posteriorly from the posterior edges of the lesions. Traditionally, in a patient with a simple rhegmatogenous detachment, lattice in the fellow eye may be observed unless an additional risk factor is present.

Source:
1. Denniston AKO, Murray PI. 2006. *Oxford Handbook of Ophthalmology*. Oxford. Oxford University Press.

48. Answer: a

Discussion: This is a gentleman with advanced glaucoma who presents with a non ischaemic CRVO, macular oedema, and a BCVA of <6/12. As per RCOphth Retinal Vein Occlusion guidelines, viable options for treatment include intravitreal dexamethasone implant and intravitreal ranibizumab. However, because of his glaucoma, ranibizumab is likely to be the safer option

Source:
1. Royal College of Ophthalmologists Interim Guidelines on Retinal Vein Occlusion 2010

49. Answer: d

Discussion: Knapp's Transposition is meant to improve elevation, by dis-insertion of the medial and lateral recti with reattachment adjacent to the insertion of the superior rectus.

50. Answer: a

Discussion: At first glance, the location of the two breaks and absence of PVR indicates that both a non encircling buckle or a pneumatic retinopexy plus external drainage are viable treatment options. However, the presence of an incomplete PVD and the anterior location of the breaks favour a buckle.
Furthermore, the reduced rate of re-operation with scleral buckle as compared to pneumatic retinopexy may be desirable in a patient keen to avoid repeat surgery.

Source:
1. Han DP, Mohsin NC, Guse CE, Hartz A, Tarkanian CN. Comparison of pneumatic retinopexy and scleral buckling in the management of primary rhegmatogenous retinal detachment. Southern Wisconsin Pneumatic Retinopexy Study Group. *Am J Ophthalmol*. 1998 Nov;126(5):658-68.
2. Brinton DA, Wilkinson CP. 2009. *Retinal Detachment : Priniciples and Practice: Priniciples and Practice*. 3rd Ed. New York. Oxford University Press.

51. Answer: c

Discussion: Surgical correction of ectropion is always aimed at correcting the causative factors.
Wedge excisions are aimed at correcting horizontal laxity.
Inverting sutures may be considered as temporary measures for patients with minimal laxity and who want to avoid more extensive surgery. Lateral tarsal strips correct lateral canthal tendon laxity.

Source:
1. Denniston AKO, Murray PI. 2006. Oxford Handbook of Ophthalmology. Oxford. Oxford University Press.

52. Answer: d

Discussion: With a dose of >7.5mg/day for >3 months, all patients above 65 should be given treatment.
For patients below 65 years, they should be treated of they have a history of fragility fractures.

For patients below 65 years without fragility fractures, a baseline bone scan should be done. A T score of <-1.5 SD indicates a need for prophylaxis; between -1.5SD and 0 indicates a need for serial bone scans to monitor bone loss progression; > 0 indicates healthy bone density and no prophylaxis is required unless very high steroid doses are used.

Source:
1. Royal College of Physicians Guidelines on Osteoporosis Prophylaxis 2002

53. Answer: c

Discussion: People with parental responsibility allowing them to give consent for a child to undergo a procedure include:
The child's birth mother
The child's married genetic father, if married at conception or birth AND is still married to the mother at the time of consultation
The child's unmarried genetic father, if present at the birth and whose name is on the birth certificate after the 1st December 2003
Grandparents or carers with a court ordered parental responsibility order

Source:
1. www.gov.uk/parental-rights-responsibilities/what-is-parental-responsibility

54. Answer: c

Discussion: The absolute risk reduction (ARR) is the difference in risk between an experimental therapy and standard therapy/control group and is usually expressed as the
risk in the standard therapy/control group – risk in the experimental group which in this case is
12/(12+48) - 15/(15+90) = 0.2 – 0.14 = +0.06
In this case the ARR is positive, which indicates that the experimental therapy is beneficial.

55. Answer: a

Discussion: The *number needed to treat (NNT)* is calculated as 1/Absolute Risk Reduction (ARR), and is expressed as the number of patients needed to be treated to avoid a single adverse event (in this scenario, defined as developing a visual acuity of <6/60).

As absolute risk reduction (ARR) is 0.06, NNT = 1/0.06 = 17. In other words, 17 patients need to be treated with the experimental therapy in order to avoid a single patient eventually developing a visual acuity of <6/60.

56. Answer: d

Discussion: According to Royal College recommendations, Haigis or Hoffer Q may be more accurate for eyes with axial lengths of less than 22 mm.

Source:
1. Royal College of Ophthalmologists Cataract Surgery Guidelines 2010

57. Answer: b

58. Answer: d

Discussion: Expulsive haemorrhage occurs due to rupture of the short posterior ciliary arteries with resulting bleeding into the suprachoroidal space. Risk factors include atherosclerosis, intraoperative tachycardia, hypertension, glaucoma, high myopia and sudden decompression of the globe – which occurs in large incision cataract surgery, penetrating keratoplasties and posterior capsular rupture with vitreous loss.

Source:
1. Speaker MG, Guerriero PN, Met JA, Coad CT, Berger A, Marmor M. A case-control study of risk factors for intraoperative suprachoroidal expulsive hemorrhage. *Ophthalmology.* 1991 Feb;98(2):202-9; discussion 210.
2. Obuchowska I, Mariak Z. Risk factors of massive suprachoroidal hemorrhage during extracapsular cataract extraction surgery. *Eur J Ophthalmol.* 2005 Nov-Dec;15(6):712-7.

59. Answer: a

Discussion: Optic disc pits are caused by imperfect closure of the embryonic optic fissure, putting the eye at risk of a 2 layered maculopathy. Both a retinal schisis as well as a neurosensory retinal detachment may occur, causing visual loss. Investigations include optical coherence tomographic (OCT) and angiographic studies of the retina (which typically shows pooling beneath a serous retinal detachment) Optic disc pits are often unilateral.

Source:
1.Lincoff H, Lopez R, Kreissig I, Yannuzzi L, Cox M, Burton T. Retinoschisis associated with optic nerve pits. *Arch Ophthalmol* 1988; 106: 61–67.

60. Answer: c

Discussion: HSV 1 is predominantly responsible for orofacial infection and is spread via secretions whereas HSV 2 is predominantly responsible for genital disease and is spread via sexual contact. Primary HSV 1 infection manifests typically as an upper respiratory infection and/or a unilateral follicular blepharoconjunctivitis. Recurrence (manifesting either as a epithelial, stromal or endothelial keratitis, keratouveitis or posterior uveitis) is caused by reactivation of the virus within the trigeminal ganglion – there is currently no evidence to suggest that stress or various environmental agents act as triggers of recurrence. Epithelial keratitis tends to be self-limiting and resolves spontaneously within 3 weeks.

Source
1. Jain V, Pineda R. Reactivated herpetic keratitis following laser in situ keratomileusis. *J Cataract Refract Surg.* May 2009;35(5):946-8
2. Patel NN, Teng CC, Sperber LT, Dodick JM. New onset herpes simplex virus keratitis after cataract surgery. *Cornea.* Jan 2009;28(1):108-10

61. Answer: b

Discussion: PONSM is predominantly a disease of middle-aged women (median age reported to be 46 years at presentation) Progesterone receptors have been reported to be present, which may explain the accelerated growth during pregnancy. Whilst the tumors retain the ability for local invasion, they rarely do so in older patients, being more aggressive in the young – which requires early diagnosis and often, surgical intervention.
A histopathologic diagnosis is often difficult, as patients often retain relatively good vision and the need for significant amounts of tissue poses a challenge. MRI has been reported to correctly differentiate PONSM from the more benign optic nerve glioma in >90% of patients.

Source:
1. Wright JE, McNab AA, McDonald WI. Primary Optic Nerve Sheath Meningioma. *Br J Ophthal.* 1989,73, 960-966
2. Lesch KP, Fahlbusch R.Simultaneous estradiol and progesterone receptor analysis in meningiomas. *Surg Neurol* 1986;26:257-63.

62. Answer: d

Discussion: Whilst any lesion along the visual pathways may theoretically cause a defect involving the patient's right visual field, the above scenario points to clues regarding location.
Pre-geniculate and geniculate lesions often, but not invariably, give rise to optic atrophy as the ganglion cell axons terminate in the thalamus. As the patients optic discs are normal, a lesion involving the pre-geniculate visual pathways or thalamus is less likely.
Lesions of the geniculate body and internal capsule often result in weakness or loss of sensation involving the contralateral body.
A stroke involving the occipital lobe (in this case the left lobe) is most likely, as the patient has had history of stroke, no other neurologic symptoms, and normal visual acuity (as occipital infarcts often spare fixation).

Source:
1. Bogousslavsky J, Caplan L. 2001. *Stroke Syndromes.* 2nd Ed. Cambridge. Cambridge University Press

63. Answer: b

Discussion: Mucormycosis tends to occur in two main groups of people – those with uncontrolled diabetes, and those who are immunocompromised with neutropenia. Intravenous drug use, skin trauma in wooded areas, transplants patients on long term anti rejection drugs and patients with malignancies are all at risk.

Source:
1. Kontoyiannis DP, Lewis RE. Agents of mucormycosis and Entomophthoramycosis. In: Mandell GL, Bennett GE, Dolin R, eds. *Mandell,, Douglas and Bennett's Principles and Practice of Infectious Diseases.* 7th ed. Philadelphia, Pa: Churchill Livingstone; 2010:3257-69.
2. http://www.cdc.gov/fungal/mucormycosis/risk-prevention.html

64. Answer: a

Discussion: Imperforate valves of Hasna are the usual causes of obstruction. The obstruction spontaneously resolves in 90% of involved children by 1 year, and if persistent, probing may be done that carries a 90% success rate. Should probing fail, a DCR is indicated.

Source:
1. Maheshwari R. Results of Probing for Congenital Nasolacrimal Duct Obstruction in Children Older than 13 Months of Age. *Indian J Ophthalmol.* 2005:53(1)49-51

65. Answer: a

Discussion: Duane's Syndrome is isolated in 70%. It is known to be associated with Morning Glory Syndrome, Wildervank Syndrome and Goldenhar Syndrome. Type 1 is the most common, and presents with an esotropia and significantly reduced abduction. It more commonly involves the left eye.

66. Answer: d

Discussion: Rod monochromatism is an autosomal recessively inherited disorder characterised by deficiency in all three cone populations. Children present at birth with nystagmus and photophobia, visual acuity of <6/60 with no colour perception. Peripheral fields are often normal. The fundus is often normal. ERG shows an absent photopic response with a normal scotopic response.

Source:
1. Tasman W, Jaeger EA. 2013. *Duane's Ophthalmology*. Lipincott Williams & Wilkins.

67. Answer: c

Discussion: Inner and outer fields of the Hess chart are equally compressed in neurogenic strabismus while restrictive strabismus demonstrate an outer field that is compressed to a greater degree compared to the inner field.

Source:
1. Denniston AKO, Murray PI. 2nd Edition. *Oxford Handbook of Ophthalmology*. Oxford. Oxford University Press

68. Answer: b

Discussion: Surgery to repair retinal detachments caused by acute retinal necrosis is fraught with difficulty; amongst others, difficulty in identification of the primary breaks, the high incidence of anterior PVR and risk of further break formation with re-detachment all necessitate the insertion of silicone oil. Multiple procedures are often necessary, and visual outcome remains poor.

Source:
1. Ahmadieh H, Soheilian M, Azarmina M, Dehghan MH, Mashayekhi A. Surgical management of retinal detachment secondary to acute retinal necrosis: clinical features, surgical techniques, and long-term results. *Jpn J Ophthalmol*. 2003 Sep-Oct;47(5):484-91

69. Answer: c

Discussion: PCO is caused by residual LEC in the capsular bag which then migrate posteriorly to proliferate over the posterior capsule between said capsule and IOL.
Steps that can be taken to minimise PCO formation include a capsulorhexis sized slightly smaller than the lens optic, meticulous hydrodissection assisted cortical clean up, in the bag implantation, the use of truncated edged, hydrophobic, posterior angulated IOLs.

Source:
1. Pandey SK, Apple DJ, Werner L, Maloof AJ, Milverton EJ. Posterior Capsule Opacification : A Review of the Aetiopathogenesis, Experimental and Clinical Studies and Factors for Prevention. *Current Ophthalmology* 2004 vol 52 Issue:2:99-112

70. Answer: b

Discussion: This is an inferior retinal detachment involving the macula, being significantly higher temporally. According to Lincoff's rules, the primary break should be along the superior border of the higher side.

Source:
1. Ryan SJ. 2013. *Retina*. 5th Ed. Elsevier.

71. Answer: d

Discussion: Calcium deposits in band keratopathy lie at the level of Bowman's membrane. They start off as innocuous deposits at 3 and 9 o clock which then progress centrally. They are often symptomatic late in the condition as the epithelium starts to break down, causing painful erosions. The deposits stain black with Von Kossa staining.

72. Answer: b

Discussion: The White Dot Syndromes typically present with bilateral manifestations with the exception of MEWDS which is unilateral in 80% of cases

Source:
1. Jampol LM, Sieving PA, Pugh D. Multiple evanescent white dot syndrome. I. Clinical findings. *Arch Ophthalmol*. 1984;102:671-674.

73. Answer: c

Discussion: Posterior synechiae, iridoschisis and silicone oil in an aphakic eye may cause pupil block and angle closure glaucoma which may respond to a patent iridotomy. ICE syndrome, on the other hand, may present with angle closure glaucoma caused by peripheral anterior synechiae, which typically does not respond to an iridotomy.

74. Answer: b

Discussion: Of the classic stromal dystrophies, corneal grafts are often required in lattice and macular dystrophy. Granular and Avellino less commonly require grafting.
Recurrences are rare post grafting in macular dystrophy.

75. Answer: c

Discussion: Overall tumor related mortality rate is 25%. Unfavourable factors for survival include unfavourable locations (caruncle, fornices, plica, lid margins), mixed cell tumors, lymphatic invasion, multifocal growth and >4mm thickness (for tumors in unfavourable locations only)

Source:
1. Paridaens AD, McCarty AC, Minassian DC, Hungerford JL. Orbital Exenteration in 95 cases of primary conjunctival malignant melanoma. *Br J Ophthalmol.* Jul 1994;78(7):520-8

76. Answer: a

Discussion: While the gold standard for investigation of direct fistulae remains intra-arterial catheter angiography, CT or MRI may demonstrate enlarged cavernous sinuses, enlarged superior ophthalmic veins and enlarged extraocular muscles.

Source:
1. Chen CCC, Chang PCT, Shy CG, Chen WS, Hung HC. CT Angiography and MR Angiography in the Evaluation of Carotid Cavernous Sinus Fistula Prior to Embolization: A Comparison of Techniques. *AJNR 2005 26: 2349-2356*

77. Answer: d

Discussion: The commonest causes of post-traumatic keratitis include bacterial and fungal causes, notably *Pseudomonas spp* and filamentous fungi. The appearance of the corneal lesion in the above scenario suggests a fungal cause. Filamentous fungi, while being the principal causes of post-traumatic infection have a distinct geographical pattern of distribution according to family. While *Aspergillus* species are the most common fungal keratitis isolates worldwide and are seen in more northern regions, *Fusarium* spp are the most common isolates in warmer climates.

Source:
1. Rosa RH Jr, Miller D, Alfonso EC. The changing spectrum of fungal keratitis in south Florida. *Ophthalmology.* Jun 1994;101(6):1005-13
2. Srinivasan M. Fungal keratitis. *Curr Opin Ophthalmol.* Aug 2004;15(4)321-7

78. Answer: d

Discussion: Albinism refers to a heterogenous group of hereditary disorders characterized by abnormal melanin synthesis and/or distribution. The two main categories are oculocutaneous albinism (which is more common and autosomal recessively inherited) and pure ocular albinism (which is less common and x linked recessive).
Ocular features typical for both categories include photophobia, nystagmus, strabismus, iris transillumination defects and foveal hypoplasia. Visual acuity is variable and ranges from 6/12 to 3/60.

Source:
1. Dijkstal JM, Cooley SS, Holleschau AM, King RA, Summers CG. Change in Visual Acuity in Albinism in the Early School Years. *J Pediatr Ophthalmol Strabismus.* Jul 6 2011;1-6

79. Answer: c

Discussion: Coats disease is characterized by painless, retinal vascular anomalies resulting in massive sub retinal exudation. There is no racial predilection, is unilateral in 80% of cases, and affects males 3 times as often as females. As 2/3 of cases present before 10 years of age and in some instances at birth, one of the main differential diagnoses is retinoblastoma.

Source
1. Asdourian G. Vascular anomalies of the retina. In: Petman GA, Sanders DR, Goldberg MF, eds. Principles and practices of ophthalmology, vol 2. Philadelphia: WB Saunders, 1980
2. Campbell FP. Coats disease and congenital vascular retinopathy. *Trans Am Ophthalmol Soc 1976; 24:365-424*
3. Egerer I, Tasman W, Tomer TL. Coat's disease. *Arch Ophthalmol 1974 92:109-11*

80. Answer: b

Discussion: A useful entry point towards the diagnosis of a possible corneal dystrophy is the presentation. A presentation of pain, photophobia and corneal erosions points to a possible diagnoses of either EBMD, RB, Type 1 or Type 3 Lattice, Granular or Avellino.

Type 2 Lattice less commonly presents with recurrent corneal erosions.

Source:
1. Roy FH, Fraunfelder FW Jr, Fraunfelder, FT. Roy and Fraunfelder's Current Ocular Therapy. 6th Ed. 2008 Elsevier.

81. Answer: b

Discussion: Differentials for subretinal solid lesions include naevi, melanomas, melanocytomas, metastases, osteomas, haemangiomas and retinal pigment epithelial (RPE) lesions such as congenital hypertrophy of the RPE. Often, ultrasonography, fluorescein and indocyanine angiography and to a lesser extent, MRI will aid in diagnosis.

The abovementioned findings strongly suggest choroidal metastases. High internal reflectivity on ultrasonography, early hypofluorescence and lack of intralesional blood vessels help distinguish it from melanomas, which show a dual circulation and low internal reflectivity.
Haemangiomas will show early hyperfluorescence.

82. Answer: b

Discussion: >10 degrees of excyclotorsion on double Maddox rod testing suggests a bilateral 4th nerve palsy in the presence of other suggestive features.

83. Answer: c

Discussion: PMD is associated with bilateral, inferior, non-inflammatory corneal thinning. Symptoms include reducing best-corrected spectacle acuity caused by progressive irregular astigmatism. Videokeratography typically shows inferior corneal thinning and steepening, with central flattening.

Source:
1. Oie Y, Maeda N, Kosaki R, Suzaki A, Hirohara Y, Mihashi T, et al. Characteristics of ocular higher-order aberrations in patients with pellucid marginal corneal degeneration. *J Cataract Refract Surg*. Nov 2008;34(11):1928-34.

84. Answer: a

Discussion: The genetic makeup of uveal melanoma is increasingly being used as predictors of mortality. Monosomy 3 is associated with loss of the gene coding for BAP1 and is predictive of metastatic disease. Gains of chromosome 8q and 6p may increase the predictive value of monosomy 3 and are associated with a better and worse prognosis respectively.

Source:
1. Sandinha MT, Farquharson MA, McKay IC, Roberts F. Monosomy 3 Predicts Death but Not Time until Death in Choroidal Melanoma. *Investigative Ophthalmology & Visual Science*. October 2005, Vol. 46, No. 10
2. www.ocularmelanoma.org

85. Answer: d

Discussion: An ipsilateral RAPD is seen in patients with optic neuropathies, and a contralateral RAPD may be seen in patients with unilateral optic tract lesions.
Lesions of the visual pathways posterior to where the pupillary light fibres leave the tracts will not give rise to an RAPD.
Strokes involving the thalamus may affect the lateral geniculate body, temporal lobe gliomas may cause compression of the optic radiations, pituitary microadenomas do not cause chiasmal compression – these lesions therefore typically will not present with an RAPD.
Anterior cerebral artery aneurysms may cause compression of the optic nerves and give rise to an RAPD.

Source:
1. Newman NJ, Miller NR, Biousse V. 2008. *Walsh and Hoyt's Clinical Neuro-Ophthalmology: The Essentials*. 2nd Ed. Lippincott Williams & Wilkins

86. Answer: b

Discussion: Stargardt's disease presents with reduced visual acuity in childhood. Fundus albipunctatus and Oguchi Disease are congenital stationary night blindness variants and present with non-progressive nyctalopia whilst Bardet Biedl Syndrome presents with progressive nyctalopia.

Source:
1. Brodsky MC. 2010. *Pediatric Neuro-Ophthalmology*. 2nd Edition. Springer
2. www.blindness.org

87. Answer: c

Discussion: MG is characterized by the production of autoantibodies to the post synaptic AchR at the neuromuscular junction. Ocular features are the presentation in roughly 50% of patients with MG. 90% of patients with MG have ocular involvement at some point in their illness but only 16% of patients presenting with ocular features do not develop generalized weakness.

Source:
1. Shah AK. *Myasthenia Gravis*. emedicine.medscape.com

88. Answer: c

 Discussion: Reiter's predominantly affects young males and consists of a triad of non-specific urethritis, conjunctivitis and arthritis. It may be post infectious, following a bout of non-gonoccocal urethritis (chlamydial, Yersinia) or cervicitis. 12% of patients develop acute anterior uveitis.
Aortic incompetence is a rare life threatening complication

89. Answer: a

Discussion: Case control studies are retrospective, observational studies that select samples based on the absence or presence of disease. Information is then collected about risk factors.

90. Answer: b

Discussion: According to joint Royal College of Ophthalmologists and Royal College of Paediatrics and Child Health guidelines, all babies born before 27 weeks of gestation should be screened at 30-31 weeks post menstrual age; all other babies requiring screening should be screened at 4-5 weeks post natal age.

Source:
1. Joint Royal College of Ophthalmologists and Royal College of Paediatrics and Child Health Guidelines for Retinopathy of Prematurity Screening 2008

Paper 4

1. In which of the following genes are mutations associated with primary congenital glaucoma?

a. CYP1B1
b. COL2A1
c. VHL
d. GLC1A

2. Which of the following pathogens is the MOST common cause of cavernous sinus thrombosis (CST)?

a. Streptococcus pneumonia
b. Staphylococcus epidermidis
c. Staphylococcus aureus
d. Aspergillus spp.

3. Regarding prisms, which of the following statements is INCORRECT?

a. The total angle of deviation is the sum of the deviations produced at both surfaces
b. The overall direction of deviation is always towards the base of the prism
c. The true image formed by the prism is deviated towards the apex of the prism
d. When constructed into prism bars, should be held with the rear surface in the frontal plane of the patient

4. The following are examples of systemic connective tissue diseases that may affect the eye and the type of hypersensitivity reaction they represent EXCEPT?

a. SLE – Type 3 Hypersensitivity
b. Rheumatoid Arthritis – Type 3 Hypersensitivity
c. Vernal Keratoconjuntivitis – Type 4 Hypersensitivity
d. Wegener's Granulomatosis – Type 1 Hypersensitivity

5. Which of the following statements regarding genetic counseling of retinoblastoma is MOST likely to be false?

a. Children of patients with hereditary retinoblastoma carry a 50% risk of the disease
b. Siblings of children with retinoblastoma and a positive family history carry a 50% risk of the disease
c. Children of patients with bilateral retinoblastoma carry a 50% risk of the disease
d. A patient with unilateral retinoblastoma without a positive family history who wants to start a family has a 1% risk of having a child with retinoblastoma

6. Behcet's disease is MOST likely to be associated with which Human Leukocyte Antigen?

a. HLA B51
b. HLA A29
c. HLA A5
d. HLA B29

7. The following conditions are associated with microspherophakia EXCEPT?

a. Peter's Anomaly
b. Weill Marchesani
c. Hyperlysinaemia
d. Congenital cytomegalovirus infection

8. Of the following conditions, which is not a common cause of stellate keratic precipitates (KP)?

a. CMV retinitis
b. Fuchs Iridocyclitis
c. Sarcoidosis
d. Herpetic Keratouveitis

9. The following statements regarding glaucoma in Fuch's Heterochromic Iridocyclitis are FALSE except?

a. Is predominantly bilateral
b. Occurs in 75% of patients
c. There is no gender bias
d. May be caused by peripheral anterior synechiae and secondary angle closure

10. What is the risk of progression of a non-ischaemic central retinal vein occlusion (CRVO) to an ischaemic CRVO?

a. 15% in a month
b. 15% in 5 months
c. 30 % in a month
d. 30% in 5 months

11. Regarding colour vision defects in acute demyelinating optic neuritis (ADON), which statement is MOST likely to be correct?

a. Colour vision defects in ADON tend to occur late in the course of the disease
b. Almost exclusively involve the red – green system
c. The type of colour defect may be used as diagnostic criteria
d. Colour vision defects tend to persist even after full recovery of visual acuity

12. Regarding malignant melanoma of the eyelids, which statement is MOST likely to be false?

a. BRAF and NRAS mutations are associated with melanomas caused by sun exposure
b. The majority of cutaneous melanoma are thought to develop in precursor naevi
c. Ulceration is an important prognostic factor
d. The predominant mode of metastasis is lymphatic

13. Which of the following disorders of amino acid metabolism is the MOST likely diagnosis for the following scenario?

Your Nephrology colleagues refer a 6-year-old boy with reduced vision to your clinic. He has glomerulonephritis. On examination, he has bilateral cataracts and raised intraocular pressure. Both his optic discs demonstrate glaucomatous cupping. His mother says he has not yet learned to talk, and that he suffers from seizures.

a. Zellweger Syndrome
b. Homocystinuria
c. Lowe's Syndrome
d. Tyrosinaemia

14. A 20-year-old gentleman presents with a large angle esodeviation involving his right eye.
His vision is OD: Counting Fingers (CF) OS: 6/6.
The deviation is unilateral and persistent, angle is similar at distance and at near.
He has history of congenital cataract in his right eye. Surgery was done which was complicated with secondary glaucoma. His optic disc is fully cupped.

What is the most likely diagnosis?

a. Congenital Esotropia
b. Sensory Esotropia
c. Consecutive Esotropia
d. Decompensated Accommodative Esotropia

15. What is the typical location for retinal dialyses associated with ocular contusion?

a. Superotemporal
b. Inferotemporal
c. Superonasal
d. Inferonasal

16. Regarding Stickler Syndrome, choose the CORRECT statement.

a. It is predominantly autosomal recessive
b. It is an uncommon cause of retinal detachments in children
c. It demonstrates similar systemic features to Wagner disease
d. It is associated with sensorineural deafness

17. The following statements are true about investigations for infective keratitis EXCEPT?

a. Culture of choice for patients with vegetative trauma would be Lowenstein Jensen agar
b. Patients with a follicular blepharo-conjunctivitis and dendritic keratitis may be investigated with a Papinicoloau stain
c. Patients with contact lens related keratitis with radial keratoneuritis should be considered for corneal biopsies should scrapes be negative
d. Scrapes of the leading edge of an ulcer should be from the ulcer edge towards the base

18. The following statements regarding imaging of the brain and orbit are false EXCEPT?

a. An acute ischemic cerebral stroke is best detected by contrasted computed tomographic (CT) scans
b. Optic Nerve Sheath Haematomas are best detected by Magnetic Resonance Imaging (MRI) FLAIR sequences
c. In a CT for a patient with suspected orbital fractures, contrast is not essential
d. Premedication with oral steroids is essential for patients with prawn allergies undergoing contrasted CT scans

19. Which of the following methods is BEST used to estimate the visual acuity of a 9-month-old child?

a. Keeler Cards
b. Kay Pictures
c. Sheridan Gardiner Test
d. Sonksen Silver Test

20. Which of the following is likely to be TRUE regarding the side effects of ocular angiography?

a. The major allergen in shellfish and other mollusks and crustaceans is an iodinated glycosaminoglycan
b. Indocyanine is contraindicated in patients with allergy to crab
c. Fluorescein angiograms are safe in patients with shellfish allergies
d. Free iodine atoms in contrast media are the causes of hypersensitivity

21. Which of the following is the MOST likely diagnosis for the following scenario?

A 20-year-old gentleman presents with unilateral blurring of vision. On examination he has a right orange retinal lesion involving the macula with exudation. There is a feeding artery and draining vein. Fluorescein angiography shows early hyperfluorescence of the lesion and diffuse leakage.

a. Retinal Capillary Haemangioma
b. Racemose Haemangioma
c. Retinal Cavernous Haemangioma
d. Retinal Macroaneurysm

22. Which of the following statements regarding pre-operative investigations for strabismus surgery is INCORRECT?

a. A post-operative diplopia test is mandatory for all binocular patients
b. In binocular patients, the ability to attain fusion post-operatively may be tested with free prisms
c. The synotophore is able to compensate for a manifest, torsional misalignment while testing the potential for binocular single vision
d. Objective measurement tests are superior compared to subjective testing in determining the amount of deviation that should be corrected.

23. Which of the following is TRUE about immunosuppressive drugs and their side effects?

a. Azathioprine causes leucopenia in up to 70% of patients
b. Methotrexate may cause hepatotoxicity in up to 50% of patients
c. Mycophenolate has no effect on glucose levels
d. Cyclophosphamide may cause frank haematuria

24. The following statements about carbonic anhydrase inhibitors (CAI) are FALSE except?

a Topically administered dorzolamide reduces intraocular pressure by 15%
b. Dorzolamide and brinzolamide penetrate the cornea poorly
c. A significant side effect of topical dorzolamide is metabolic acidosis
d. Risk of hypokalemia is increased on concurrent diuretic therapy

25. The commonest organism causing post-operative endophthalmitis (POE) is

a. Coagulase positive Staphylococci
b. Coagulase negative Staphylococci
c. Beta Haemolytic Streptococci
d. Pseudomonas aeruginosa

26. Which of the following pathogens is MOST often the causative factor of bacterial conjunctivitis in kindergarten–age children?

a. Neisseria gonorrhea
b. Pseudomonas aeruginosa
c. Haemophilus influenza
d. Group C Streptococci

27. Which of the following features is LEAST likely to be a feature of adenoviral keratoconjunctivitis?

a. Subconjunctival haemorrhage
b. Preauricular Lymphadenopathy
c. Bulbar follicles
d. Membrane formation

28. The following statements regarding glaucoma in herpetic eye disease are likely to be false EXCEPT?

a. Is more commonly seen in association with concurrent epithelial keratitis as compared to stromal keratitis
b. Is predominantly an angle closure glaucoma
c. Is often seen during the first presentation of herpetic eye disease
d. Glaucoma filtration devices carry a higher success rate as compared to augmented trabeculectomies

29. What is the MOST likely diagnosis for the following scenario?

A 59-year-old lady with hypertension presents with dilated veins in the supero-temporal quadrant of her left eye associated with intraretinal haemorrhages. The macula is oedematous,

a. Branch Retinal Vein Occlusion
b. Hemi retinal Vein Occlusion
c. Branch Retinal Artery Occlusion
d. Central Retinal Artery Occlusion

30. A 7-year-old girl presents with progressive, painless, bilateral blurring of vision. He visual acuity is 6/96 in both eyes. On examination, she has widespread bone spicule pigmentary changes involving both eyes with macular scarring. She wears hearing aids and has been deaf from birth. What is the MOST likely diagnosis?

a. Bardet - Biedl Syndrome
b. Usher Syndrome
c. Kearn - Sayre Syndrome
d. Refsum Disease

31. Which of the following disorders is the MOST likely diagnosis for the following scenario?

A 7-year-old girl presents with bilateral blurring of vision. Her visual acuity is OD: 6/60 OS: 6/96 There is a round, well-demarcated scar involving both maculas. The rest of the fundus is normal. EOG shows an Arden ratio of 1.2

a. Stargardt's Disease
b. Best's Vitelliform Dystrophy
c. Adult Foveolar Vitelliform Dystrophy
d. Toxoplasmosis Retinitis

32. According to Royal College guidelines for the management of Central Retinal Vein Occlusion, the following statements are false EXCEPT?

a. Patients with ischaemic CRVO should be monitored every 2 months initially to look for evidence of anterior segment neovascularisation
b. A presenting corrected visual acuity of 6/36 is associated with an increased risk of neovascular glaucoma
c. Patients presenting with a corrected visual acuity of 6/15 and macular oedema should be considered for intravitreal dexamethasone (Ozurdex) implant
d. A central foveal thickness of 200μm is an indication for treatment with Ozurdex

33. Which of the following statements regarding Child Protection is MOST likely to be false?

a. Obtaining parental consent is imperative prior to examination in all cases of suspected child abuse
b. If a parent refuses to consent to examination of a child in a case of suspected abuse, emergency legal action may be initiated by the police
c. The Named Doctor (ND) or Named Nurse (NN) are the personnel in each Trust who are responsible for dissemination of child protection guidelines
d. Each Local Health Authority (LHA) should have a Designated Doctor (DD) or Designated Nurse (DN)

34. Regarding Phase 3 trials, which of the following statements is TRUE?

a. Involves determination of drug safety and efficacy of therapeutic doses in human subjects
b. Involves testing of a drug on healthy volunteers to determine the therapeutic dose
c. Involves determination of pharmacokinetics and pharmacodynamics, particularly, oral bio-availability in human subjects
d. Involves determination of efficacy and toxicity in non human subjects

35. Which of the following statements regarding Paget's disease is MOST likely to be incorrect?

a. Characterised by excessive bone remodelling
b. Is not sight threatening
c. Associated with malignancies
d. Prognosis is good if treatment is started before major bone changes have occurred

36. Posterior ischemic optic neuropathy (PION) differs from non-arteritic anterior ischemic optic neuropathy (NA - AION) in the following ways EXCEPT?

a. Optic disc is normal in the acute phase in PION
b. There is often a normal angiogram in the acute phase in PION
c. There is usually a normal optic disc in PION as opposed to a small or absent cup in AION
d. Central scotomas are a common occurrence in AION as compared to PION

37. Regarding the prognosis for Acute Demyelinating Optic Neuritis, which of the following statements is MOST likely to be false?

a. Up to 90% of patients eventually recover to attain a visual acuity of 6/9 or better in the affected eye
b. Female sex is a risk factor for eventually developing multiple sclerosis
c. The length of optic nerve enhancement on magnetic resonance imaging (MRI) correlates with visual prognosis
d. The 10- year recurrence rate (in either eye) after an initial episode is 15%

38. Regarding treatment options and prognosis of Leber's Hereditary Optic Neuropathy, which of the following statements is MOST likely to be false?

a. Visual acuity usually deteriorates to 6/60 or worse, then stabilizes
b. The second eye is usually involved within 2 months
c. There is often some degree of spontaneous recovery
d. There is no effective treatment

39. Thiamine deficiency is MOST likely to be associated with

a. Bariatric surgery
b. Chronic Alcoholism
c. Gastric Carcinomas
d. Hyperemesis Gravidarum

40. Regarding Kaposi Sarcoma of the head and neck, which statement is MOST likely to be false?

a. Occurs exclusively in immunocompromised patients
b. Is considered an acquired immunodeficiency syndrome (AIDS) defining condition
c. Human Herpesvirus 8 seroconversion is a risk factor
d. Cutaneous involvement is typically non-pruritic

41. Which of the following disorders is LEAST likely to cause a cicatricial entropion?

a. Steven Johnson's Syndrome
b. Trachoma
c. Pseudomembranous conjunctivitis
d. Partial thickness laceration of the lower lid involving only the skin

42. In which of the following conditions is acute proptosis LEAST likely to respond to a canthotomy and cantholysis?

a. Thyroid Eye Disease
b. Retrobulbar haemorrhage
c. Orbital Emphysema
d. Optic nerve glioma

43. Which of the following statements regarding senile retinoschisis is MOST likely to be incorrect?

a. Is often bilateral
b. Tends to involve the infero-nasal quadrant
c. Is associated with hyperopia
d. Rarely involves the posterior pole

44. Select the most appropriate management option for the following scenario

A 58-year-old lady with hypertension and diabetes presents with sudden onset blurring of vision in her right eye for a week. Her corrected visual acuity is Counting Fingers (CF) in that eye, with a brisk relative afferent pupillary defect. Fundus examination reveals dilated veins and haemorrhages in all quadrants, cotton wool spots and macular oedema.

a. Follow up monthly with serial acuity checks, gonioscopy, and dilated fundoscopy
b. Follow up 2 monthly with serial acuity checks, gonioscopy, and dilated fundoscopy
c. Intravitreal ranibizumab
d. Intravitreal dexamethasone implant

45. The following procedures affect the direction of pull of the operated muscle EXCEPT?

a. Hummelsheim
b. Inverse Knapp
c. Faden Procedure
d. Harada Ito

46. A 39-year-old gentleman with Type 1 diabetes presents with tractional macular detachment in his left eye. He is also myopic and has extensive lattice degeneration extending from 11 to 2 o clock.

What is the most likely surgical treatment option for him?

a. Pars plana vitrectomy and endolaser retinopexy
b. Pars plana vitrectomy, delamination, endolaser retinopexy and silicone oil tamponade
c. Scleral buckling and gas tamponade
d. Circumferential scleral buckle

47. Which of the following grafts is LEAST likely to be used for reconstructing the posterior lamellae of the lower lid?

a. Buccal Mucosa
b. Hard Palate
c. Donor Sclera
d. Auricular Cartilage

48. A neonate born premature at 26 weeks with a birth weight of 890 grams has been scheduled for her first retinopathy of prematurity (ROP) screening. On examination she has Stage 1 ROP with Plus disease in Zone 1 in her right eye and Stage 3 ROP without Plus disease in Zone 1 in her left eye. What is the most appropriate course of action?

a. Observe and follow up weekly
b. Observe and follow up two weekly
c. Discharge
d. Treat

49. The following are non-parametric tests EXCEPT?

a. Kruskal Wallis
b. One Way ANOVA
c. Wilcoxon Signed Ranks Test
d. Mann Whitney

50. Regarding malignant hyperthermia (MH), which of the following statements is MOST likely to be false?

a. Characterised by a massive exothermic response
b. Typically results in acute renal failure
c. History of uncomplicated general anaesthesia (GA) reliably excludes risk of malignant hyperthermia
d. Treatment is with intravenous dantrolene

51. Which of the following biometric readings should prompt a repeat measurement?

a. Axial length of <22.00 mm
b. Mean K reading of >46.5D
c. Delta K of >2.5D
d. Difference in axial length of >0.3 mm

52. Regarding the Kayser Fleischer ring in Wilson's Disease, choose the CORRECT statement.

a. Caused by copper deposition at the level of Bowman's Membrane
b. The ring is often visible to the naked eye at an early stage of formation
c. Are present in individuals with Wilson's Disease who are otherwise asymptomatic
d. They initially appear at the superior margin of the cornea

53. Which of the following statements regarding features of optic nerve hypoplasia is MOST likely to be false?

a. May be bilateral in up to 90% of cases
b. Visual acuity is generally very poor
c. A double ring sign may be present
d. The disc – foveal distance is increased

54. The following statements regarding herpes zoster ophthalmicus (HZO) are true EXCEPT?

a. Is caused by a double stranded DNA virus
b. Pain is a prominent feature
c. The vesicles typically crust and heal over a period of 3 months
d. A vesicular rash involving the tip of the nose is a strong predictor of corneal denervation and ocular inflammation

55. What is the percentage of patients with optic nerve glioma (ONG) that have Neurofibromatosis Type 1 (NF-1)?

a. 10%
b. 20%
c. 30%
d. 40%

56. Which of the following is NOT a feature of a unilateral optic tract lesion?

a. Abnormal visual acuity
b. Normal light brightness
c. Contralateral Relative Afferent Pupillary Defect (RAPD)
d. Bow tie atrophy of the contralateral optic disc

57. Which of the following statements regarding rhabdomyosarcoma is MOST likely to be true?

a. Is the most common solid extracranial tumour of childhood
b. Is derived from pluripotent stem cells
c. Head and neck rhabdomyosarcoma are usually of the alveolar histopathologic variant
d. Radiotherapy is not a viable treatment option

58. Which of the following disorders of lipid metabolism is MOST likely to describe the following scenario?

A 30-year-old gentleman with mitral valve prolapse, progressive renal impairment and a painful red rash extending from his knees to his umbilicus, presents to your clinic. He has whorl like corneal epithelial deposits.

a. Tay Sachs Disease
b. Fabry's Disease
c. Krabbe's Disease
d. Farber's Disease

59. Which of the following conditions is LEAST likely to present with an esodeviation?

a. Thyroid eye disease with restrictive myopathy
b. Traumatic fracture of the ethmoid sinus
c. Raised Intracranial Pressure
d. Type 2 Duane's Syndrome (Huber Classification)

60. Which of the following genes is NOT implicated in rod monochromatism?

a. CNGA3
b. CNGB3
c. COL2A1
d. GNAT2

61. What is the inheritance pattern of Type 1 Congenital Fibrosis of the Extraocular Muscles (CFEOM)?

a. Autosomal Dominant
b. Autosomal Recessive
c. X Linked
d. Mitochondrial

62. Regarding retinal detachments in association with cytomegalovirus (CMV) retinitis, which of the following statements is MOST likely to be false?

a. Is caused by breaks in necrotic retina
b. The rate of retinal detachment is related to the type of intravenous therapy
c. Risk is not increased by myopia
d. There is no difference in long term rates of retinal detachment between patients receiving ganciclovir implants and patients receiving intravenous therapy

63. What is the molecular weight of sodium hyaluronate?

a. 500 KDa
b. 50 KDa
c. 5 KDa
d. 5000 KDa

64. A 39-year-old gentleman has a shallow rhegmatogenous retinal detachment in his left eye. It extends from 4 to 8 o clock.

Where is the likeliest site of the primary break?

a. Above the horizontal meridian
b. 6 o clock
c. 12 o clock
d. None of the above

65. Regarding inheritance patterns of corneal dystrophies, which of the following statements is FALSE?

a. Epithelial Basement Membrane Dystrophy (EBMD) is Autosomal Dominantly (AD) inherited
b. Thiel Benke (Bowman's Layer 2 Corneal Dystrophy) is Autosomal Recessively (AR) inherited
c. Reis Buckler Dystrophy (Bowman's Layer 1 Corneal Dystrophy) is Autosomal Dominantly inherited
d. Macular Dystrophy is Autosomal Recessively inherited

66. Which of the following corneal dystrophies is NOT associated with mutations of the BIGH3 gene?

a. Granular Dystrophy
b. Avellino Dystrophy
c. Macular Dystrophy
d. Type 1 Lattice Dystrophy

67. Which of the following statements regarding conjunctival naevi is MOST likely to be true?

a. Most often involves the superior limbus
b. Are usually acquired
c. Cysts are seldom seen
d. Are usually amelanotic at birth

68. The following conditions are differential diagnoses for poorly defined orbital lesions on orbital imaging EXCEPT?

a. Orbital lymphoma
b. Orbital Cellulitis
c. Cavernous Haemangioma
d. Adenoid cystic carcinoma of the lacrimal gland

69. When is interstitial keratitis (IK) in association with acquired syphilis MOST commonly seen?

a. Before 10 years of age
b. Early teens
c. 3rd to 5th decade of life
d. After 60 years of age

70. Systemic complications associated with albinism include the following EXCEPT?

a. Progressive pulmonary fibrosis
b. Cardiomyopathy
c. Thyroid malignancy
d. Cutaneous malignancies

71. What is the proportion of people of African descent with Sickle-Cell Haemoglobin C (HbSC) disease?

a. 8.5%
b. 2.5%
c. 0.15%
d. 0.2%

72. Which of the following peripheral corneal diseases occurs in a non-inflamed eye?

a. Marginal keratitis
b. Pellucid marginal degeneration
c. Mooren's
d. Peripheral Ulcerative Keratitis

73. According to the following investigations, what is the MOST likely diagnosis?

A 50-year-old lady with a unilateral, flat, orange – yellow, well demarcated choroidal lesion in both eyes with pseudopod – like edges. Ultrasound shows highly reflective anterior borders with orbital shadowing. CT scan shows bilateral dense lesions at the level of the choroid. MRI reveals hyperintense T1 signals and hypointense T2 signals.

a. Choroidal Metastasis
b. Choroidal Osteoma
c. Choroidal Melanoma
d. Choroidal Haemangioma

74. What is the MOST likely diagnosis for the following scenario?

A 17-year-old girl with chronic earache presents with sudden onset binocular diplopia. On examination, she is febrile, drowsy, has difficulty abducting her right eye and complains of right hemifacial pain.

a. Ophthalmoplegic Migraine
b. Gradenigo Syndrome
c. Tolosa Hunt Syndrome
d. 6th nerve palsy

75. Pertaining to Cytomegalovirus (CMV), which statement is MOST likely to be correct?

a. Causes the majority of Human Immunodeficiency Virus (HIV)/Acquired Immune Deficiency Syndrome (AIDS) related infectious retinopathies
b. Prior to the advent of Highly Active Anti Retroviral Therapy (HAART), the incidence of CMV retinitis in patients with CD4 counts of less than 50/ul was 30%
c. It is a single stranded DNA virus of the Herpes group
d. Is characterised by vasculitis and perivascular retinitis

76. Which of the following statements is regarding Adie's Tonic Pupil is MOST likely to be correct?

a. Tends to occur in young men
b. Tonicity of the near response is an early feature
c. The pupil remains dilated for the rest of the natural history of the disease
d. Up - regulation of muscarinic receptors on the iris sphincter occurs

77. In the CRYO-ROP Study, what was the rate of reduction (in terms of %) of unfavourable structural outcome in treated eyes vs. non-treated eyes?

a. 56%
b. 46%
c. 36%
d. 66%

78. Which of the following does NOT cause hyperfluorescence during fluorescein angiography?

a. Dry Age Related Macular Degeneration (AMD)
b. Irvine Gass syndrome
c. Active Vogt Koyanagi Harada (VKH) Syndrome
d. Choroidal Naevi

79. Which of the following statements is MOST likely to be true concerning the relationships between HLA B27 related diseases and anterior uveitis?

a. 10% of HLA B27 positive patients develop uveitis
b. 55% of patients with acute anterior uveitis are HLA B27 positive
c. The recurrence rate of acute anterior uveitis is similar in patients who are HLA B27 positive as compared to HLA B27 negative patients
d. 1% of patients with Reiter's syndrome develop anterior uveitis

80. Regarding auto-antibodies in Myasthenia Gravis (MG), which of the following statements is MOST likely to be true?

a. An acetylcholine receptor (AchR) antibody test has a less likely to be positive in patients with purely systemic myasthenia
b. The absence of striated muscle antibodies (SM) should prompt a search for thymoma
c. The overwhelming majority of seronegative MG patients test positive for muscle specific kinase antibodies (MuSK)
d. Antinuclear antibody testing is indicated

81. Regarding Magnetic Resonance Angiography (MRA), which statement is MOST likely to be false?

a. Non-invasive method of imaging carotid and vertebro-basilar systems
b. Is gadolinium enhanced
c. Does not detect thrombosed aneurysms
d. Is based on motion sensitivity of MRI

82. Regarding psoriasis, choose the INCORRECT statement.

a. Psoriatic arthropathy is usually benign
b. 20% of patients with purely cutaneous psoriasis develop anterior uveitis
c. 20% of patients with cutaneous psoriasis develop psoriatic arthropathy
d. The uveitis associated with psoriasis typically presents in both eyes simultaneously

83. Regarding randomisation, which randomisation method is LEAST likely to be acceptable?

a. According to age or date of birth
b. Block
c. Stratified
d. Adaptive

84. Regarding screening for visual field defects in patients on Vigabatrin, which of the following statements is MOST likely to be false?

a. Screening for as yet asymptomatic visual field defects allows for cessation of the drug
b. Should be initiated in patients old enough to comply with a visual field assessment
c. A Humphrey threshold 24-2 test is the screening method of choice
d. As long as no field defects are present, patients should be screened 6 monthly for 5 years, and annually after

85. Which of the following pigmented ocular lesions is MOST likely to be bilateral?

a. Naevus of Ota
b. Benign Conjunctival Melanosis
c. Conjunctival Naevus
d. Primary Acquired Melanosis (PAM)

86. The following statements regarding internuclear ophthalmoplegia (INO) are TRUE except?

a. Caused by lesions involving the medial pons or midbrain
b. A lesion on the right side of the brainstem will cause loss of adduction of the right eye
c. Conjugate gaze towards the side of the lesion is affected
d. May be detected by optokinetic drum testing

87. All of the following may cause and artificially short axial length on an A - scan EXCEPT?

a. Contact A - scans
b. Beam misalignment
c. Asteroid Hyalosis
d. Pseudophakic settings in a phakic eye

88. Which of the following White Dot Syndromes is associated with HLA A29?

a. Serpiginous Choroidopathy
b. Birdshot Choroidopathy
c. Multifocal Choroiditis and Panuveitis (MCP)
d. Acute Posterior Multifocal Placoid Pigment Epitheliopathy (APMPPE)

89. Choose the INCORRECT statement regarding OCT

a. Employs low coherence light with the application of interferometry
b. Differences in tissue characteristics alter the signal intensity
c. Increased visualization of the choroid may be seen in patients with macular scarring
d. Is degraded by the presence of asteroid hyalosis

90. Which of the following is MOST likely to be true concerning methicillin resistant Staphylococcus aureus (MRSA) screening?

a. Plays an important role in ophthalmology as it is a significant cause of post operative endophthalmitis (POE)
b. All patients undergoing day case surgery are required to be screened by the Department of Health
c. De-colonization regimes have been proven to be highly effective in reducing the rates of endophthalmitis caused by MRSA
d. MRSA colonization of the nares is a contraindication to cataract surgery

Paper 4 Answers and Discussion

1. Answer: a

Discussion: Mutations in the gene CYP1B1 are associated with primary congenital glaucoma, whereas COL2A1 is the gene coding for Type 2 Collagen and is implicated in the pathogenesis of Stickler Syndrome. VHL is implicated in Von Hippel Lindau and GLC1A is implicated in primary open angle glaucoma.

Source:
1. Stone EM, Fingert JH, Alwad WLM, Nguyen TD, Polansky JR, Sunden SLF et al. Identification of a Gene That Causes Primary Open Angle Glaucoma. *Science 31 January 1997:* Vol. 275 no. 5300 *pp. 668-670*

2. Answer: c

Discussion: *Staphylococcus aureus* remains the cause in up to 70% of CST cases, and therefore should be considered when starting intravenous antibiotic therapy. First line agents include a 3rd or 4th generation cephalosporin for gram-negative coverage plus intravenous cloxacillin/vancomycin for *Staph aureus* coverage.

Source:
1. Sharma R. Cavernous Sinus Thrombosis. *emedicine.medscape.com*

3. Answer: c

Discussion: The total angle of deviation is the sum of the deviations produced at both surfaces. The overall direction of deviation of the incident ray is always towards the base of the prism, thus forming a real image that is deviated towards the base, and a virtual image that is deviated towards the apex.
The minimum angle of deviation is produced when the amount of deviation is equal at both surfaces. The designated power of prisms incorporated into a prism bar is calculated according to their respective minimum angles of deviation, and thus should be held with their rear surface in the frontal plane of the patient.

Source:
1. 2009-2010. *Clinical Optics.* American Academy of Ophthalmology

4. Answer: d

Discussion: Wegener's granulomatosis demonstrates features of Types 2,3,4 Hypersensitivity.
VKC demonstrates mainly a Type 1 and a prominent Type 4 reaction. SLE and Rheumatoid Arthritis demonstrate features of a Type 3 reaction

5. Answer: d

Discussion: Children of patients with hereditary retinoblastoma have a 50% risk of having retinoblastoma. Virtually all children with bilateral retinoblastoma regardless of family history and children with a family history regardless of laterality have hereditary disease.

Virtually all children with unilateral disease without a family history have non-hereditary retinoblastoma. Their offspring carry an 8% risk of having the disease.

Risk of the following persons to have a child with retinoblastoma		
	Parent of Patient	Patient
Positive Family History	50%	50%
No Family History		
Bilateral	2%	50%
Unilateral	1%	8%

Source:
1. Draper GJ, Sanders BM, Brownbill PA, Hawkins MM. Patterns of risk of hereditary retinoblastoma and applications to genetic counselling. *Br J Cancer.* 1992 Jul;66(1):211-9.
2. Wills Eye Institute. Ocular Oncology Service. www.retinoblastomainfo.com

6. Answer: a

Discussion: Behcet's was strongly associated with HLA B51 or B5 in a recent systematic review

Source:
1. de Menthon M, Lavalley MP, Maldini C, Guillevin L, Mahr A. HLA-B51/B5 and the risk of Behçet's disease: a systematic review and meta-analysis of case-control genetic association studies. *Arthritis Rheum*. 2009 Oct 15;61(10):1287-96.

7. Answer: d

Discussion: Microspherophakia refers to an abnormally small and spherical lens. It is usually inherited via an autosomal recessive pattern in association with Weill Marchesani syndrome, but may be associated with other systemic conditions. It arises from mutations in the LTPB2 gene, which causes abnormal weakening of the zonules. Patients are highly myopic.
Pupil block glaucoma due to the abnormally long antero-posterior diameter of the lens may occur. The block may be broken with cycloplegics that tighten the zonules and shorten the antero-posterior diameter. Alternatively, a laser iridotomy may be preferred.
(Causes: **F**riday **P**artying and **M**errymaking **W**ill **H**ave **A** **C**ost = Familial Microspherophakia Peter's Anomaly Marfan's Weill Marchesani, Hyperlysinaemia Alport syndrome Congenital Rubella)

Source:
1. Nirankari, M.S.; Maudgal, M.C. (1959). "Microphakia". British Journal of Ophthalmology (43): 314–316.
☐ Kumar et al. (October 2010). "A homozygous mutation in LTBP2 causes isolated microsperophakia". Human Genetics **128** (4): 365–371.
3. American Academy of Ophthalmology One Network. One.AAO.org

8. Answer: c

Discussion: Stellate KPs are fine, star shaped, and distributed diffusely over the corneal endothelium. They typically occur in herpetic keratouveitis, and may be seen in Fuchs Iridocyclitis and CMV retinitis.
Sarcoidosis typically presents with granulomatous inflammation and mutton fat KPs.

Source:
1. Jap A, Chee SP. Viral Anterior Uveitis. *Curr Opin Ophthalmol*. 2011 Nov;22(6):483-8.

9. Answer: c

Discussion: FHIC is unilateral in 90% of patients, has no gender bias, and causes glaucoma in approximately 15% of patients. It is an open angle glaucoma, and synechiae, in contrast to herpetic uveitis and iridocorneal endothelial (ICE) syndrome, are usually not seen. Augmented trabeculectomies have been reported to carry a success rate of 72% at 2 years.

Source:
1. La Hey E, de Vries J, Langerhorst CT, Baarsma GS, Kijlstra A. Treatment and prognosis of secondary glaucoma in Fuchs Heterochromic Iridocyclitis. *Am J Ophthalmol*. 1993 Sep 15;116(3):327-40.

10. Answer: b

Discussion: The risk of progression of an initially non-ischaemic CRVO to an ischaemic CRVO is 15% in 5 months and 34% by 3 years.

Source:
1. Central Vein Occlusion Study Group. Central Vein Occlusion Study. *Ophthalmology* 1995;102:1425-1433

11. Answer: d

Discussion: Colour vision defects observed in the Optic Neuritis Treatment Trial included red-green, blue-yellow, and non-specific defects. The defects tended to be very severe at the onset of the demyelinating episode. The proportion of blue yellow defects in the acute phase of the episode tended to be higher with a noticeable shift in type to predominantly red – green defects at 6 months that persisted even after complete recovery of visual acuity. Therefore, as types of colour vision defects vary widely in patients with ADON, it is advisable that color vision type not be used as a diagnostic criteria. (regardless of what Kollner's rule states!)

Source:
1. Schneck ME, Haegerstrom-Portnoy G. Color Vision Type and Spatial Acuity in the Optic Neuritis Treatment Trial. *Invest Ophthalmol Vis Sci*. 1997;38:2278-2289.

12. Answer: b

Discussion: While the causes for cutaneous melanoma are multifactorial, BRAF and NRAS mutations are associated with melanomas caused by sun exposure. The majority of melanomas (70%) arise de novo. The important prognostic factors include tumor thickness, ulceration and lymph node involvement, of which the latter is the most important.

Source:
1. Lee JH, Choi JW, Kim YS. Frequencies of BRAF and NRAS mutations are different in histologic types and sites of origin of cutaneous melanoma: a meta analysis. *Br J Dermatol.* Dec 16:2010

13. Answer: c

Discussion: The classic triad of cataracts with or without glaucoma, mental handicap/seizures and chronic, progressive renal impairment should alert one to the possibility of Lowe's Syndrome (Oculocerebrorenal Syndrome)

14. Answer: b

Discussion: The above differentials of *a persistent, unilateral esodeviation* with poor vision would include 2 main differentials;
An intermittent esophoria which has long since decompensated with resulting amblyopia OR
Sensory esotropia, which is caused by poor vision in the deviating eye.

In this scenario, the latter is more likely as there is evidence of organic pathology likely preceding the deviation

15. Answer: b

Discussion: Inferotemporal retinal dialyses are pathognomonic of an ocular contusion.

Source:
1. Ferenc Kuhn. 2008. *Ocular Traumatology.* Springer.

16. Answer: d

Discussion: Stickler's syndrome is an autosomal dominant hereditary vitreoretinopathy. It is the most common cause of retinal detachment in children. It has characteristic ocular and systemic features including myopia, cataract, retinal detachment, deafness and Pierre Robin sequence. Wagner is similar to Stickler but lacks systemic features. Furthermore, Wagner is not associated with an increased risk of retinal detachments.

Source:
1. Liberfarb RM, Levy HP, Rose PS et al. The Stickler syndrome: genotype/phenotype correlation in 10 families with Stickler syndrome resulting from seven mutations in the type II
2. Ryan, SJ. 2006. *Retina.* 4th Ed. Philadelphia, PA. Mosby Elsevier.

17. Answer: a

Discussion: Scrapes should be performed with either a no. 15 blade, Kimura spatula or 25 gauge needle. Base of the ulcer and leading edge should be scraped, from non-involved to involved cornea.
Patients with vegetative trauma and keratitis should be investigated thoroughly for a fungal cause, for which Saboraud agar is the culture media of choice.
A follicular blepharo-conjunctivitis and dendritic keratitis are signs of herpes simplex keratitis which may be investigated with a Papinicoloau stain, Giemsa stain or swabs for PCR.
Patients with suspected acanthamoeba keratitis should have cornea biopsies performed should initial swabs be negative.

18. Answer: c

Discussion: Acute ischemic strokes may be detected with diffusion weighted MRI scans within minutes of onset. Initial imaging for a patient with orbital trauma would typically be a non-contrasted CT scan that is able to detect orbital fractures, of particular importance, sphenoid fractures. Other pathology detectable by CT would include nerve sheath haematomas, retrobulbar haematomas as well as metallic foreign bodies. MRI FLAIR sequence is meant to detect paraventricular abnormalities. Shellfish allergies are not related in any way to allergies to contrast medium.

Source:
1. American College of Radiology – www.acr.org

19. Answer: a

Discussion: Visual acuity tests for children are usually chosen according to age of the child. For infants, preferential looking tests are best used e.g Keeler or Teller cards. Kay pictures may be used for preverbal or verbal children who are able to match. Sheridan Gardiner and Sonksen silver utilize letter optotypes typically meant for children aged 3 and above.

20. Answer: c

Discussion: The major allergen in shellfish and mollusks are the proteins tropomyosin, and less commonly, parvalbumin. This is the reason why patients who are allergic to one type of seafood are often allergic to another species.
The major contrast agents in ocular angiography, indocyanine green and fluorescein are totally unrelated to tropomyosin and hence, are safe in patients with seafood allergies.

There is significant myth surrounding 'iodine' allergies. Medical myth tends to hold that IgE mediated sensitivity to free iodine atoms is the main cause of any adverse reactions.
In actual fact, radio contrasted media causes non-IgE related reactions such as burning or pain secondary to either direct cellular toxicity or a hyperosmolar preparation. True IgE anaphylaxis to iodinated contrast media is thought to be related to the molecular structure of the preparation, and not the iodine.

Sources:
1. American Academy of Allergy, Asthma and Immunology –www.aaaai.org
2. Schabelman E, Witting M. The relationship of radiocontrast, iodine, and seafood allergies: a medical myth exposed. *J Emerg Med* 2010 Nov;39(5):701-7
3. Huang SW. Seafood and iodine: an analysis of a medical myth. *Allergy Asthma Proc* 2005;26:468-9.

21. Answer: a

Discussion: The findings above point to a diagnosis of a retinal capillary haemangioma, which may be associated with underlying Von Hippel Lindau disease. The lesions are typically situated between arteries and veins.and may threaten vision by causing macular exudation or exudative retinal detachment. Fluorescein angiogram shows filling of the feeding artery, followed by rapid, complete hyperfluorescence of the lesion, filling of the vein and diffuse late leakage.

In contrast, cavernous haemangiomas may show slow, incomplete filling with a fluid meniscus and no leakage while racemose haemangiomas show rapid filling with no leak.

22. Answer: a

Discussion: The post-operative diplopia test is mandatory for all non-binocular patients in the presence of a manifest squint. The method involves gradually overcorrecting the angle of deviation to move the deviated image over the retina till it moves out of the suppression scotoma and the patient experiences diplopia.

Source:
1. Kushner BJ. Intractable diplopia after squint surgery in adults. *Arch Ophthalmol.* 2002 Nov;120(11):1498-504.

23. Answer: d

Discussion: Azathioprine may cause leucopenia in up to 30% of patients. While methotrexate may cause hepatotoxicity, incidence is less than 10%. Mycophenolate may cause hyperglycaemia

Cylophosphamide has been known to cause haemorrhagic cystitis.
24. Answer: c

Discussion: CAI may be administered topically or systemically.
Topical agents include dorzolamide and brinzolamide. These agents penetrate the cornea well, and avoid the side effects associated with systemically administered CAI. Common side effects include burning and stinging.
Systemic CAI include acetazolamide and methazolamide. Side effects include metabolic acidosis and hypokalemia, especially in patients with concurrent diuretics or steroids.

25. Answer: b

Discussion: The commonest isolates are *Staphylococcus epidermidis* with an estimated prevalence of up to 77%. Infection with exotoxin producing *Staphylococci* species may be particularly severe, often resulting in poorer visual outcome.

Source:
1. ESCRS Guidelines for Prevention and Treatment of Endophthalmitis following cataract surgery. 2013

26. Answer: c

Discussion: Acute bacterial conjunctivitis in children under 6 is most often caused by *Haemophilus*. Other usual suspects include *Streptococcus pneumoniae*, *Staphylococcus aureus* and *Staphylococcus epidermidis.*

Source:
1. Wald ER. Periorbital and orbital infections. Pediatr Rev. 2004;25:312-319

27. Answer: c

Discussion: Adenoviral conjunctivitis is the commonest cause of infective conjunctivitis, particularly amongst school going children. Features include variable laterality, pre-auricular lymphadenopathy, serous discharge, chemosis, inferior tarsal follicles, membrane or pseudomembrane formation, haemorrhages as well as punctate/subepithelial keratitis.

Bulbar follicles are suggestive of chlamydial inclusion conjunctivitis

Source:
1. American Academy of Ophthalmology Corneal/External Disease Panel. Preferred Practice Pattern: Conjunctivitis. San Francisco, Ca: AAO; 2003.

28. Answer: d

Discussion: Glaucoma in association with herpetic eye disease almost never presents in patients who either present with epithelial involvement or who present with the initial episode of herpetic eye disease.

Literature suggests that glaucoma is only seen in patients with recurrent disease. The mechanism is predominantly an open angle glaucoma due to clogging of the trabecular meshwork by inflammatory cells and debris as well as inflammation of the meshwork itself. Intraocular pressure may initially be controlled with antiviral therapy in conjunction with steroids. For patients whose glaucoma is uncontrolled by maximum medical therapy, surgery with glaucoma devices have been shown to carry a success rate of over 90% at 1-2 years as compared to 50-70% with augmented trabeculectomies.

Source:
1. Falcon MG, Williams HP. Herpes simplex keratouveitis and glaucoma. *Trans Ophthal Soc* UK. 1978;98:101-104.
2. Patitsas CJ, Rockwood EJ, Meisler DM, et al. Glaucoma filtering surgery with postoperative 5-fluorouracil in patients with intraocular inflammatory disease. *Ophthalmology.* 1992;99:594-599.
3. Ceballos EM, Beck AD, Lynn MJ. Trabeculectomy with antiproliferative agents in uveitic glaucoma. *J Glaucoma.* 2002;11:189-196.3.
4. Da Mata A, Burk SE, Netland PA, et al. Management of uveitic glaucoma with Ahmed Glaucoma Valve implantation. *Ophthalmology.* 1999;106:2168-2172.
5. Ceballos EM, Parrish RK, Schiffman JC. Outcome of Baerveldt glaucoma drainage implants for the treatment of uveitic glaucoma. *Ophthalmology.* 2002;109:2256-2260.

29. Answer: a

Discussion: This is most likely to be a branch retinal vein occlusion as only the supero – temporal veins are dilated.

30. Answer: b

Discussion: Hearing loss in association with retinitis pigmentosa is seen most commonly in Usher's syndrome. The hearing loss in Usher's is often congenital, and profound. Apart from Usher's, hearing loss in association with retinitis pigmentosa is also seen in Refsum's Disease, Waardenburg Syndrome and Alport Syndrome.

31. Answer: b

Discussion: This is most likely to be a diagnosis of Best's Vitelliform Dystrophy. Differential diagnoses for bilateral macular scarring in a young patient include congenital toxoplasma infection and myopic macular degeneration. However, the reduced Arden ratio points to a diagnosis of Best's Disease. Adult Foveolar Vitelliform Dystrophy may present similarly, but occurs in adulthood.

32. Answer: c

Discussion: The management of CRVO would largely depend on whether it is ischemic or non ischemic. A brisk RAPD and a vision of <6/96 is highly suggestive of ischemia. In these cases, no treatment is indicated, and the patient should be followed up monthly initially to look for evidence of anterior segment new vessels. Patients presenting with non-ischemic CRVO and macular oedema, with a visual acuity of 6/12 or less and a central foveal thickness of more than 250μm should be offered ozurdex implant.

Source:
1. Royal College of Ophthalmologists Interim Guidelines on Retinal Vein Occlusion 2010

33. Answer: a

Discussion: NDs and NNs are Trust designated key personnel who are trained in managing suspected cases of child abuse and have responsibility for local dissemination of Child Protection Guidelines. DDs and DNs are LHA designated personnel who are available for consultation should the local Trust have neither a ND nor NN.
Prior to examination, parental or caretaker consent should be sought. However, should the child demonstrate sufficient understanding, the child's consent alone may suffice.
In the face of refusal of consent by child or parents/caretakers, emergency legal action may need to be considered – which can be initiated by either Social Services or the Police.

Source:
1. Royal College of Ophthalmologists Guidelines on Child Protection. *Ophthalmology Child Abuse Working Party in discussion with The Royal College of Paediatrics and Child Health Standing Committee on Child Protection.*

34. Answer: a

Discussion: Phase 3 trials are 'pre-approval' trials that aim to determine the efficacy and safety of the therapeutic dose of a particular drug in human subjects.

35. Answer: b

Discussion: Paget's Disease is caused by excessive bone remodelling which may cause optic canal constriction and compressive optic neuropathy. Prognosis is generally good if treatment is started before major bone changes have occurred. Pagetoid sarcomas may occur which carry a poor prognosis.

Source:
1. Eretto P, Krohel GB, Shihab ZM, Wallach S, Hay P. Optic neuropathy in Paget's disease. *Am J Ophthalmol.*1984 Apr;97(4):505-10.

36. Answer: d

Discussion: PION differs from AION in that the optic disc and angiograms are normal in the acute phase, there is absence of a 'disc at risk', and central scotomas are more common in PION as opposed to altitudinal scotomas in NA – AION.

Source:
1. Hayreh SS. Posterior ischaemic optic neuropathy: clinical features, pathogenesis, and management. *Eye* (2004) 18, 1188–1206

37. Answer: d

Discussion: After an initial demyelinating episode, 90% of patients recover to attain a Snellen acuity of 6/9 or better. However, abnormalities of low contrast acuity and colour vision defects tend to persist, as do visual evoked potential abnormalities. Poorer visual prognosis is associated with increased length of nerve enhancement on MRI.
The 10-year recurrence rate in either eye after an initial episode is 35% as per findings in the Optic Neuritis Treatment Trial (ONTT).
Risk factors for developing multiple sclerosis include female sex, presence of white matter lesions on MRI and oligoclonal bands in the cerebrospinal fluid.

Source:
1. Beck, Roy W., et al. "Visual function more than 10 years after optic neuritis: experience of the optic neuritis treatment trial." *American journal of ophthalmology* 137.1 (2004): 77-83.

38. Answer: c

Discussion: The second eye is often involved within 2 months. Visual acuity usually stabilizes at 6/60 or less. Spontaneous recovery is rare, and when it does occur, it occurs in association with the less common 14484 point mutation. Raxone, a Coenzyme Q10 analogue, has not been proven to be effective in improving the visual prognosis of LHON and has recently been rejected for approval by the European Medicines Agency.

Source:
1. www.ema.europa.eu

39. Answer: b

Discussion: The commonest cause of thiamine deficiency and resulting Wernicke's encephalopathy in the developing world is malnutrition associated with chronic alcohol abuse.

40. Answer: a

Discussion: Kaposi sarcoma are spindle cell tumors originating from endothelial cells. The spectrum of disease ranges from isolated mucocutaneous involvement to extensive organ involvement. The disease is seen in 4 settings; as part of advanced HIV where it is considered an AIDS defining illness, sporadic Kaposi, as part of immunocompromise in transplant patients as well as being endemic in Africa. In HIV related Kaposi's sarcoma, prior infection by Human Herpes Virus 8 is an independent risk factor for development of the disease. The tumors present typically as palpable papules or nodules that are typically non-pruritic. Highly Active Antiretroviral Therapy (HAART) may be considered as monotherapy for mucocutaneous lesions which are usually localised.

Source:
1. Rose LJ. *Kaposi Sarcoma.* medscape.emedicine.com
2. Cattelan AM, Calabro ML, De Rossi A, et al. Long-term clinical outcome of AIDS-related Kaposi's sarcoma during highly active antiretroviral therapy. *Int J Oncol.* Sep 2005;27(3):779-85.

41. Answer: d

Discussion: Cicatricial entropion arise from inflammation and subsequent scarring of the tarsal conjunctiva and subconjunctival tissues, with buckling of the tarsal plates and internal rotation of the lid margins. They are sight-threatening conditions, causing chronic irritation of the corneal surface with subsequent scarring and vascularisation. Partial thickness lacerations of the lower lids may cause scarring of the skin and external rotation of the lid margins, resulting instead in a cicatricial ectropion.

42. Answer: d

Discussion: Canthotomies and cantholyses are indicated in cases of Orbital Compartment Syndrome (OCS), a sight threatening disorder characterised by increasing pressure within the orbit, leading to vascular compromise involving the optic nerve.

Common causes are retrobulbar haemorrhages caused by trauma or surgery, orbital emphysema, cellulitis, oedema, and thyroid eye disease. By lysing the canthal tendons, the lids are able to move forward, allowing dissipation of intraorbital pressure.

Optic nerve gliomas are less likely to fill the orbital space, and instead, may cause optic neuropathy by direct compression of the optic nerve.

Source:
1. Lima V, Burt B, Leibovitch I, Prabakharan V, Goldberg RA, Selva D. Major Review: Orbital Compartment Syndrome – The Ophthalmic Surgical Emergency. *Surv Ophthalmol.* 54(4):July-August 2009
2. Condon JR, Rose FC. Optic Nerve Glioma. *Br J Ophthalmol.* 1967 October; 51(10): 703–706.

43. Answer: b

Discussion: Senile retinoschisis has been reported to be bilateral in 82%, to maximally involve the infero-temporal quadrant, and to be associated with hyperopia.

It rarely involves the posterior pole and rarely progresses to retinal detachment.

Source:
1. Byer NE. Clinical Study of Senile Retinoschisis. *Arch Ophthalmol.* 1968;79(1):36-44.

44. Answer: a

Discussion: This is a lady who presents with an ischaemic CRVO. Appropriate management in this scenario would be to avoid treatment as per RCOphth Retinal Vein Occlusion guidelines, and to follow-up monthly for up to 6 months to detect the presence of anterior segment neovascularization.

Source:
1. Royal College of Ophthalmologists Interim Guidelines for Retinal Vein Occlusion 2010

45. Answer: c

Discussion: Faden Procedure or myopexy, involves postequatorial fixation of the muscle to the sclera with a non absorbable suture. This weakens the pull of the muscle without altering the direction of pull.

Source:
1. Wright KW, Spiegel PH. 2nd Edition. *Pediatric Ophthalmology and Strabismus.* New York. Springer.

46. Answer: b

Discussion: Vitrectomy and delamination of the fibrovascular membranes to eliminate traction on the macula is the most accepted form of treatment. There is usually an option to insert a long lasting intraocular tamponade such as C3F8 gas or silicone oil in order to lend support to the detached retina, but there is some evidence that suggests that if there are no associated breaks, a tamponade may not necessarily be inserted. However, given the fact that the patient is myopic, and does have risk factors for posterior vitreous detachment related tears, silicone oil tamponade may be prudent.

Source:
1. Tao Y. Long term results of vitrectomy without endotamponade in patients with proliferative diabetic retinopathy. *Retina.* 2010;30(3):447-451
2. Garg A, Alio JL. 2010. *Surgical Techniques in Ophthalmology: Retina and Vitreous Surgery.* New Delhi. Jaypee Brothers Medical Publishers.

47. Answer: a

Discussion: Due to the effects of gravity, thin oral mucosa is less ideal for reconstruction of the posterior lamellae of the lower lid. More rigid material is required.

Source:
1. Tyers AG, Collin JRO. 2008. *Colour Atlas of Ophthalmic Plastic Surgery. 3rd Ed.* Elsevier

48. Answer: d

Discussion: The child mentioned above has pre-threshold disease in both eyes. According to UK ROP guidelines 2008, she should be treated with laser photocoagulation.

Source:
1. United Kingdom Retinopathy of Prematurity Screening Guidelines 2008

49. Answer: b

Discussion: Of the mentioned tests, the One Way Anova is the only parametric test, and is used to compare means between two sets of parametric (randomized, normally distributed) data sets.

50. Answer: c

Discussion: MH is a life threatening disorder characterised by altered skeletal muscle sarcoplasmic reticulum calcium channel kinetics. Massive calcium ion influx is triggered by exposure to succinylcholine and volatile anaesthetics, causing rigidity, rhabdomyolysis and a massive exothermic response. Multiorgan failure and disseminated intravascular coagulation is the main cause of death. A history of uncomplicated GA does not reliably exclude risk. Currently, the only method of reliably excluding MH is by the caffeine halothane contracture test (CHCT) that should be considered in patients with any family history or anaesthetic history suggestive of MH. Treatment is with intravenous dantrolene.

Source:
1. Metterlein T, Hartung , Schuster F, Roewer N, Anetseder M. Sevoflurane as a potential replacement for halothane in diagnostic testing for malignant hyperthermia susceptibility: results of a preliminary study. *Minerva Anestesiol.* Aug 2011;77(8):768-73

51. Answer: c

Discussion: Axial lengths of <21.20 mm or >26.60 mm, mean K readings of >47 D or <41D, a difference in inter-ocular mean K reading of 1.0 D or more and a delta K of >2.5D should prompt a repeat measurement.

Source:
1. Royal College of Ophthalmologists Cataract Surgery Guidelines 2010

52. Answer: d

Discussion: KF rings are caused by copper and sulfur deposits within Descemet's membrane. They appear initially at the superior margin, then inferior, and finally circumferentially. Well-developed rings may be seen with the naked eye, and early rings may be detected with the slit lamp or gonioscopy. The rings are classically seen in symptomatic Wilson's disease (and are invariably seen in patients with neurological features) but may also be seen in patients with chronic cholestatic disorders e.g primary biliary cirrhosis.

Source:
1. Schilsky ML. Wilson disease: Current status and the future. *Biochimie.* Jul 30 2009

53. Answer: b

Discussion: Optic nerve hypoplasia has been reported to be bilateral in 56-92% of cases. Visual function is widely variable, and ranges from no symptoms to severe visual loss. The disc is small, and may be pale and grey. A double ring sign may be present. Disc foveal distance is increased.

54. Answer: c

Discussion: HZO is caused by the varicella zoster virus, a double stranded DNA virus of the Herpesviridae family. After primary infection (manifesting as chicken pox) the virus becomes dormant within sensory ganglia throughout the body. Reactivation of the virus within the trigeminal sensory ganglia causes a vesicular rash involving the dermatomes supplied by the trigeminal nerve. Hutchison's sign, referring to involvement of the side or tip of the nose by the rash, is a strong predictor of associated ocular inflammation and corneal denervation. The rash usually crusts and heals within 2-6 weeks.

Source:
1. Hutchinson J. A clinical report on herpes zoster ophthalmicus (shingles affecting the forehead and nose). *Trans Am Ophthalmol Soc* 1942;40:390-439.

55. Answer: c

Discussion: 30% of patients with ONG have NF-1, which carries a better prognosis.

Source:
1. Orbit, Eyelids and Lacrimal System, Section 7. Basic and Clinical Science Course, AAO, 2011-2012
2. Jack J Kanski, Brad Bowling. 2011. *Clinical Ophthalmology: A systemic approach*. 7th ed. Elsevier Saunders

56. Answer: a

Discussion: Features of isolated, unilateral optic tract lesions include normal visual acuity (assuming the chiasm is spared), normal colour vision and light brightness, the presence of a contralateral RAPD, bow tie atrophy of the contralateral optic disc and diffuse pallor of the ipsilateral disc. The classic visual field defect is a contralateral homonymous hemianopia.

Source:
1. Newman NJ, Miller NR, Biousse V. 2008. *Walsh and Hoyt's Clinical Neuro-Ophthalmology: The Essentials*. 2nd Ed. Lippincott Williams & Wilkins

57. Answer: b

Discussion: Rhabdomyosarcoma is the 3rd most common solid extracranial tumour of childhood, after Wilm's tumour and nephroblastoma. It is thought to derive from pluripotent stem cells committed to the skeletal muscle lineage. Head and neck rhabdomyosarcoma is the most common site, and is predominantly comprised of the embryonal histopathologic variant (which is thought to be a favourable prognostic factor). Treatment options depend on the staging, and consist of local control methods with adjuvant chemotherapy and radiotherapy.

Source:
1. Anderson JR, Link M, Qualman S et al. Improved outcome for patients with embryonal histology but not alveolar histology RMS: results from Intergroup Rhabdomyosarcoma Study-IV. *Proc Am Soc Clin Oncol* 1998 ;17 :526a
2. Dagher R, Helman L. Rhabdomyosarcoma: An Overview. *The Oncologist* February 1999 vol. 4 no. 1 34-44

58. Answer: b

Discussion: Fabry's disease is an X linked disorder of lipid metabolism caused by a deficiency of alpha-galactosidase. Patients present early with angiokeratomas, which is a painful red rash that is most dense from the knees to the umbilicus and vortex keratopathy. In their 30s, they start to develop cardiac and renal complications, most often mitral valve prolapse and progressive renal impairment.

Source:
1. *www.fabry.org*

59. Answer: d

Discussion: Type 2 Duane's Syndrome presents with an exodeviation, mildly reduced abduction, and severely reduced adduction.

Source:
1. Wright KW, Spiegel PH. 2nd Edition. *Pediatric Ophthalmology and Strabismus*. New York. Springer.

60. Answer: c

Discussion: COL2A1 is one of the genes coding for collagen 2, and is implicated in 80% of patients with Stickler syndrome. CNGA3, CNGB3 and GNAT2 code for proteins involved in the cone phototransduction pathway.

Source:
1. Brodsky MC. 2010. *Pediatric Neuro-Ophthalmology*. 2nd Edition. Springer

61. Answer: a

Discussion: Classic CFEOM (Type 1) which presents with bilateral ptosis and congenital restrictive ophthalmoplegia is an autosomal dominant disorder.

Source:
1. Engle EC et al. Congenital fibrosis of the extraocular muscles (autosomal dominant congenital external ophthalmoplegia): genetic homogeneity, linkage refinement, and physical mapping on chromosome 12. *Am J Hum Genet*. 1995 Nov;57(5):1086-94.

62. Answer: b

Discussion: Retinal detachments in association with CMV retinitis are caused by breaks within or at the border of healed, necrotic lesions. Rates of detachment are not related to type of intravenous therapy or refractive error, and while rates of detachment have been found to be higher within the first 2 months in patients receiving ganciclovir implants as compared to patients receiving intravenous therapy, in the long term there is no statistical difference in rate.

Source:
1. Studies of Ocular Complications of AIDS (SOCA) Research Group in Collaboration with the AIDS Clinical Trials Group (ACTG). Rhegmatogenous retinal detachment in patients with cytomegalovirus retinitis. Foscarnet-ganciclovir Cytomegalovirus Retinitis Trial. *Am J Ophthalmol* 1997; 124:61-70
2. Ryan, SJ. 2006. *Retina*. 4th Ed. Philadelphia, PA. Mosby Elsevier.

63. Answer: a

Discussion: Sodium hyaluronate consists of the sodium salt of hyaluronan, a visco elastic polymer found in both the vitreous and aqueous humour. The molecular weight of sodium hyaluronate is in the region of 500-750 KDa.

64. Answer: b

Discussion: An inferior retinal detachment equal in height on both sides would point to the primary break being at 6 o clock.

Source:
1. Ryan SJ. 2013. *Retina*. 5th Ed. Elsevier.

65. Answer: b

Discussion: A simple way to remember inheritance patterns is to group the classic corneal dystrophies according to location.
Of the classic epithelial dystrophies (EBMD, Reis Buckler, Thiel Benke and Meesmans) – AD inheritance is the norm
Of the classic stromal dystrophies (Granular, Lattice, Avellino and Macular) – AD inheritance is the norm EXCEPT for Macular Dystrophy which is AR
Of the classic endothelial dystrophies (Fuchs, Posterior Polymorphous Dystrophy and Congenital Hereditary Endothelial Dystrophy (CHED) – AD inheritance is the norm except for CHED which is AR

66. Answer: c

Discussion: BIGH3 mutation is associated with Type 1 Lattice, Granular and Avellino dystrophy

Source:
1.. Mashima Y, Yamamoto S, Inoue Y, Yamada M, Konishi M, Watanabe H, Maeda N, Shimomura Y, Kinoshita S. Association of autosomal dominantly inherited corneal dystrophies with BIGH3 gene mutations in Japan. *Am J Ophthalmol*. 2000 Oct;130(4):516-7

67. Answer: d

Discussion: Conjunctival nevi are congenital lesions that are usually amelanotic at birth and gain pigmentation at puberty. They are most often seen in young, Caucasian patients. A large case series (410 patients) demonstrated frequent involvement of the nasal (46%) or temporal (44%) limbus, presence of cysts (65%) and feeder vessels (33%).

Source:
1. Zembowicz A, Mandal RV, Choopong P. (*2010*) Melanocytic Lesions of the Conjunctiva. *Archives of Pathology & Laboratory Medicine*: December 2010, Vol. 134, No. 12, pp. 1785-1792.
2. Shields CL, Fasiuddin AF, Mashayekhi A, Shields JA. Conjunctival nevi: clinical features and natural course in 410 consecutive patients. . *Arch Ophthalmol*. 2004 Feb;122(2):167-75

68. Answer: c

Discussion: Orbital mass lesions can be characterized according whether their edges are well or poorly defined as seen on computed tomographic (CT) scans or magnetic resonance imaging (MRI) scans. Poorly defined lesions have hazy margins that tend to indicate infiltration or inflammation. Cavernous haemangioma does neither and thus has well defined edges.

69. Answer: c

Discussion: IK is classically associated with syphilis infection and less commonly, with systemic vasculitides as well as herpetic keratitis. *Mycobacterium tuberculosis* and *Mycobacterium leprae* are also known causes.
IK associated with congenital syphilitic infection usually presents within the first decade of life; acquired IK within the 3rd-5th decade.

Source:
1. Majmudar PA. Interstitial Keratitis. *emedicine.medscape.com*

70. Answer: c

Discussion: Hermansky-Pudlak syndrome is associated with platelet dysfunction, progressive lung fibrosis, inflammatory bowel disease, cardiomyopathy and renal impairment. Chediak-Higashi syndrome is associated with lymphoid malignancies and recurrent infection due to abnormal leukocyte chemotaxis.

Source:
1. Demirkiran O, Utku T, Urkmez S, Dikmen Y. Chediak Higashi syndrome in the intensive care unit. *Paediatr Anaesth*. Aug 2004;14(8):685-8
2. Carter BW. HErmansky Pudlak syndrome complicated by pulmonary fibrosis. *Proc (Bayl Univ Med Cent)*. Jan 2012;25(1):76-7

71. Answer: d

Discussion: Amongst the various sickle cell haemoglobinopathies, the variants most likely to manifest with retinal features are HbSC and sickle cell thalassaemia (HbSThal).
Of the proportion of North American citizens of African descent, the proportion of those with HbSC and HbSThal has been estimated to be 0.2% and 0.03% respectively.

Source:
1. Gagliano DA, Jampol L, Rabb M. Sickle cell disease. In: Tasman WS, Jaeger E meds, Duane's clinical ophthalmology. Philadelphia: Lipincott Raven, 1996; vol 3:1-40
2. Jandl J. Blood. In: Pathophysiology. Cambridge: Blackwell Scientific Publications, 1991.
3. Steinberg M. Management of sickle cell disease. *N Engl J Med* 1999; 340:1021-1030

72. Answer: b

Discussion: A reasonable entry point to formulation of a diagnosis in a patient with peripheral corneal thinning would be laterality as well as whether or not inflammation is present.
The above conditions tend to be bilateral and simultaneous, with the exception of marginal keratitis (which may be asymmetrical) and are associated with inflammation except in the case of Pellucid degeneration which occurs in a white, quiet eye.

73. Answer: b

Discussion: Differentials for subretinal solid lesions include naevi, melanomas, melanocytomas, metastases, osteomas, haemangiomas and retinal pigment epithelial (RPE) lesions such as congenital hypertrophy of the RPE. Often, ultrasonography, fluorescein and indocyanine angiography and to a lesser extent, MRI will aid in diagnosis.

The findings above are typical of choroidal osteomas. They may be bilateral in a quarter of patients and the cause is yet unknown. Chronic choroiditis has been postulated. Ultrasonography shows a very reflective anterior border due to density of the bone and orbital shadowing from attenuation of the sound wave causing loss of available signals from tissue deep to the osteoma. T1 MRI reveals hyperintense signals in relation to water due to fat deposition between the bony trabecular spaces and T2 reveals hypointense signals in relation to water because of paucity of water.

Source:
1. Depotter P, Shields J.A, Shields C.L, Rao V.M. Magnetic resonance imaging in choroidal osteoma. *Retina*. 11(2): 221-223

74. Answer: b

Discussion: Gradenigo syndrome is a rare, potentially lethal cause of 6th nerve palsy. It is caused by apical petrositis and extradural inflammation (often secondary to chronic otitis media) that involves both the 6th nerve as well as the trigeminal nerve and is characterized by 6th nerve palsy and facial pain. Potential life threatening complications include intracranial abscess and meningitis.

Source:
1. Motamed M, Kalan A. Gradenigo Syndrome. *Postgrad Med J* 2000;**76**:559–560

75. Answer: b

Discussion: CMV is a double stranded DNA virus of the Herpes group. While the prevalence of CMV retinitis in individuals with HIV/AIDS remains higher than that of other infectious retinopathies, the prevalence of HIV related microvasculopathy remains the highest of the HIV/AIDS related infectious retinopathies at roughly 50%, and may affect the vision if the perifoveal capillaries are affected. Prior to 1995-1997, the prevalence of CMV retinitis in individuals with HIV/AIDS was roughly 30%. With the advent of HAART in 1995, prevalence has dropped drastically.

Sources:
1. www.who.int/bulletin/archives/79(3)208.pd

76. Answer: d

Discussion: Adie's tonic pupil is unilateral in 80% and affects young healthy women. It is thought to be due to viral denervation involving the ciliary ganglion. Patients present with anisocoria that is worse in the light and a fixed dilated pupil that is non reactive to both light or attempted near response in the acute phase. Tonicity of the near response is seen in the late stage.
Vermiform movements of the sphincter may be seen on slit lamp examination. Because of upregulation of muscarinic receptors on the pupil sphincter, dilation occurs with dilute pilocarpine 0.125%.

77. Answer: c

Discussion: The rate reduction of an unfavourable structural outcome (Stage 4 ROP or worse) at one year post treatment in the CRYO-ROP study was 35.8% in treated eyes vs. non-treated eyes.

Source:
1. Cryotherapy for Retinopathy of Prematuriy Cooperative Group. Multicenter trial of cryotherapy for retinopathy of prematurity. One year outcome – structure and function. *Arch Ophthalmol* 1990;108:1408-1416

78. Answer: d

Discussion: Choroidal naevi are a cause of hypofluorescence due to a masking effect. Dry AMD causes a window defect due to retinal and retinal pigment epithelium atrophy. Irvine Gass presents with a petalloid hyperfluorescent pattern and active VKH typically presents with pooling caused by multifocal exudative detachments.

79. Answer: b

Discussion: Overall, 1% of HLA B27 positive patients develop acute anterior uveitis (AAU), and 55% of patients presenting with AAU are HLA B27+, rising to about 70% in patients with recurrent episodes.

84% of HLA B27 positive patients who develop AAU have other related systemic disease.
20-30% of patients with ankylosing spondilitis and 12-17% of patients with Reiter's syndrome develop anterior uveitis.

Source:
1. Sheehan, N.J. The Ramifications of HLA B27. *J R Soc Med.*2004 January. 97(1): 10-14

80. Answer: d

Discussion: AchR antibody testing has a sensitivity of 90% in systemic MG and 50-70% in ocular MG. False negative rates are thus lower in systemic MG. The presence of SM antibodies suggests the presence of a thymoma, especially in younger patients. 50% of seronegative MG patients (negative for AchR) test positive for MuSK antibodies which usually indicates a distinct subgroup of MG with more pronounced bulbar symptoms and a poorer response to Ach-esterase inhibitors. ANA testing is indicated to look for associated SLE or RA.

Source:
1. Padua L, Stalberg E, LoMonaco M, Evoli A, Batocchi A, Tonali P. SFEMG in ocular myasthenia gravis diagnosis. *Clin Neurophysiol*. Jul 2000;111(7):1203-7.
2. Stickler DE, Massey JM, Sanders DB. MuSK-antibody positive myasthenia gravis: clinical and electrodiagnostic patterns. *Clin Neurophysiol*. Sep 2005;116(9):2065-8.
3. Pasnoor M, Wolfe GI, Nations S, et al. Clinical findings in MuSK-antibody positive myasthenia gravis: a U.S. experience. *Muscle Nerve*. Mar 2010;41(3):370-4.
4. Sanders DB, El-Salem K, Massey JM, McConville J, Vincent A. Clinical aspects of MuSK antibody positive seronegative MG. *Neurology*. Jun 24 2003;60(12):1978-80

81. Answer: b

Discussion: MRA is non invasive, and images blood flow based on motion sensitivity of MRI. Intra and extracranial carotid as well as the vertebro - basilar systems can be imaged. It is non-contrasted and is unable to detect thrombosed aneurysms.

82. Answer: b

Discussion: 20% of patients with cutaneous psoriasis develop psoriatic arthropathy, which is typically benign. A minority of patients does however develop arthritis mutilans which is destructive.
20% of patients with psoriatic arthropathy develop anterior uveitis, which is atypical to that associated with other HLA B27 related disease in that it may present bilaterally.

83. Answer: a

Discussion: Randomisation according to age or date of birth confers some degree of predictability and subsequently, causes selection bias. Therefore these are generally not acceptable methods of randomization.

Source:
1. Wang D, Bakhai A. 2006. *Clinical Trials: A Practical Guide to Design, Analysis, and Reporting*. Remedica.

84. Answer: c

Discussion: Either Humphrey or Goldmann perimetry may be utilized, however, the sensitivity of Humphrey screening may be better. Screening should be supra-threshold strategy to at least 45 degrees of eccentricity, for example, a Humphrey Supra-threshold 120 degree full field test.

Source:
1. Royal College of Ophthalmologists Guidelines on Screening for Side Effects of Vigabatrin. March 2008

85. Answer: b

Discussion: Benign conjunctival melanosis is typically bilateral.
PAM, naevi and naevus of ota (which is a variant of oculodermal melanosis) are typically unilateral.

86. Answer: c

Discussion: Lesions involving the medical longitudinal fasciculus, which runs the length of the midbrain and pons along the midline, cause INO. It is characterised by loss of adduction of the ipsilateral eye. Conjugate gaze towards the side of the lesion is preserved. Unless the lesion involves the dorsal midbrain, convergence is often intact. Mild cases may be detected by observation of saccadic slowing of the eye on the same side as the lesion while rotating an optokinetic drum towards the side of the lesion.

Source:

1. Brazis PW, Masdeu JC, Biller J. 2011. *Localisation in Clinical Neurology.* 6th Ed. Philadelphia, PA, USA. Lippincott Williams and Wilkins

87. Answer: d

Discussion: Corneal compression in contact scans, beam misalignment and asteroid hyalosis all may cause an artificially short axial length. Phakic velocity settings in a pseudophakic eye may also lead to an artificially short axial length as sound is significantly faster in both PMMA and Acrylic as compared to the crystalline lens.

Source:
1. Waldron RG. *A-Scan Biometry.* emedicine.medscape.com

88. Answer: b

Discussion: Birdshot Choroidopathy is associated with HLA A29 in 90% of patients.

Source:
1. Gasch AT, Smith JA, Whitcup SM. Birdshot retinochoroidopathy. *Br j Ophthalmol.* 1999;83:241-249.

89. Answer: d

Discussion: OCT employs low coherence light with the application of interferometry.
Differences in characteristics of different layers alter the reflected light signal that is converted to a color coded scale. Degeneration of RPE increases visualization of the choroid. Asteroid hyalosis does not degrade the OCT image and is useful in patients with poor visualization of the fundus due to asteroid hyalosis.

Source:
1. Hwang JC. et al. Optical Coherence Tomography in Asteroid Hyalosis. *Retina.* 2006 Jul–Aug; 26(6): 661–665.

90. Answer: d

Discussion: MRSA is an infrequent cause of POE. Neither screening nor de-colonization protocols have been shown to be effective in reducing the rates of POE. Current Department of Health guidelines stipulate that day case Ophthalmology patients do not need to be screened. All admitted patients do need to be screened. MRSA colonization of the nares is not a contraindication for cataract surgery unless the periocular region is found to be colonized as well.

Source:
1. Duerden B. Inspector of Microbiology and Infection Control, the Department of Health. Letter to Royal College of Ophthalmologists. 2008 Jan
2. Beasley C, Flory D. MRSA screening – operational guidance. Gateway ref: 10324. Department of Health; 2008.
3. Porter LF, Khan RU, Hannan A, Kelly SP. MRSA and Cataract Surgery – Reflections for Practice. *Clin Ophthalmol.* 2010; 4: 1223–1227

Paper 5

1. Which of the following inheritance patterns BEST describes the inheritance pattern of Sturge Weber Syndrome?

a. A 3-year-old boy presents with a port wine stain and secondary open angle glaucoma. He is the 3rd of 4 siblings. He has an elder brother with a similar condition. His father has the same condition.

b. A 3-year-old boy presents with a port wine stain and secondary open angle glaucoma. He is the 6th of 8 siblings. He has an elder brother with a similar condition. Both his parents are unaffected.

c. A 3-year-old boy presents with a port wine stain and secondary open angle glaucoma. He is the 3rd of 4 siblings. His parents and siblings are unaffected.

d. A 3-year-old boy presents with a port wine stain and secondary open angle glaucoma. He is the 3rd of 4 siblings. Both his parents are unaffected. He has two sisters who are unaffected, and one brother who is affected. One of his maternal uncles has a similar condition.

2. A 23-year-old gentleman is admitted for severe facial burns causing thermal ocular injury, severe eyelid scarring and lagophthalmos. On the 5th day of admission, he develops a spiking fever. Which of the following pathogens is MOST likely to be responsible?

a. Staphylococcus epidermidis
b. Staphylococcus aureus
c. Pseudomonas aeruginosa
d. Enterococci spp.

3. The following statements regarding binocular single vision are true EXCEPT?

a. Corresponding retinal areas are areas in both eyes that are stimulated simultaneously by the same visual object
b. The horopter is a 3 dimensional construct
c. The images of all points within Panum's area fall on disparate or non- corresponding retinal elements and are seen as separate points
d. The stereo Fly test is a test of gross stereopsis

4. Touton giant cells are seen in

a. Xanthogranuloma
b. Sarcoidosis
c. Tuberculosis
d. Giant Cell Arteritis

5. Which of the following scenarios BEST describes the mode of inheritance of Leber's Hereditary Optic Neuropathy?

a. An 11-year-old boy presents with sudden onset, unilateral blurring of vision in his right eye followed by similar symptoms in his left eye 6 weeks later. He has 2 sisters and one brother. One of his sisters has a similar ocular history. His father has had similar symptoms in his youth.

b. An 11-year-old boy presents with sudden onset, unilateral blurring of vision in his right eye followed by similar symptoms in his left eye 6 weeks later. He has 2 sisters and one brother, all of whom are not affected. His parents are unaffected.

c. An 11-year-old boy presents with sudden onset, unilateral blurring of vision in his right eye followed by similar symptoms in his left eye 6 weeks later. He has 2 sisters and one brother, all of whom have similar symptoms. His mother and maternal siblings had similar symptoms in their early teens.

d. An 11-year-old boy presents with sudden onset, unilateral blurring of vision in his right eye followed by similar symptoms in his left eye 6 weeks later. He has 2 sisters and one brother. His brother has similar symptoms. His parents are unaffected, but he has 2 maternal uncles who have similar symptoms.

6. Which of the following conditions is NOT X-Linked Recessive?

a. Nettleship Falls Albinism
b. Lowe's Syndrome
c. Alport Syndrome
d. Homocystinuria

7. Regarding the direction of lens subluxation, which of the following statement pairs is MOST likely to be false?

a. Marfan's Syndrome – superotemporal lens subluxation
b. Familial Ectopia Lentis – superotemporal lens subluxation
c. Homocystinuria – superotemporal lens subluxation
d. Weill Marchesani Syndrome – anteroinferior lens subluxation

8. Which of the following statements is LEAST likely to be true regarding adenoviral keratitis?

a. Pharyngoconjunctival fever rarely presents with subepithelial infiltrates
b. Epidemic Keratoconjunctivitis tends to present with a severe prodrome
c. Pre-auricular lymphadenopathy is typical
d. Oral acyclovir is not effective

9. The following statements regarding glaucoma risk in congenital disorders are TRUE except?

a. 75% in Aniridia
b. 50% in Axenfeld Rieger Syndrome
c. 90% in Sturge Weber Syndrome
d. 30% at 10 years in patients who have had cataract surgery for congenital cataract before the age of 9 months

10. The following are features suggestive of an ischaemic central retinal vein occlusion (CRVO) EXCEPT?

a. Visual acuity of <6/60
b. Relative Afferent Pupillary Defect (RAPD)
c. Electronegative electroretinogram
d. Presence of 2 cotton wool spots

11. Regarding non-arteritic anterior ischemic optic neuropathy (NA-AION), which of the following statements is MOST likely to be false?

a. Is the most common cause of acute optic neuropathy in patients over 50 years of age
b. Visual loss may be secondary to nocturnal hypotension
c. Classic findings include sectoral oedema of the inferior half of the disc
d. Is typically painless

12. Regarding the prognosis of cutaneous basal cell carcinoma (BCC), which of the following statements is MOST likely to be true?

a. The prognosis is generally poor
b. The incidence of metastatic BCC is in the region of 1%
c. Cure rates with treatment are in the region of 95%
d. There is a 5% chance of developing other BCCs within 3 years

13. Which of the following intrauterine infections is MOST likely to be the culprit in the following scenario?

A 1-week-old baby girl is referred to your Paediatric clinic for an eye assessment. She has hydrocephalus and intracranial calcifications. On examination, she has bilateral macular scars. A TORCH screen is pending.

a. Rubella
b. Cytomegalovirus
c. Toxoplasma
d. Herpes Simplex

14. The following statements are TRUE regarding micro-esotropias except?

a. Usually not more than 2-3 prism dioptres
b. If present, amblyopia is often mild
c. Positive 4 dioptre prism test
d. Reduced binocular function

15. Regarding traumatic macular holes, which of the following statements is MOST likely to be false?

a. Typically occurs 6 months after trauma
b. Most often associated with ocular contusions
c. Spontaneous closure rates are higher compared to that of idiopathic macular holes
d. Vitrectomy, internal limiting membrane (ILM) peel and gas tamponade is highly successful in inducing closure

16. Refractive errors in association with different hereditary vitreoretinopathies are as follows EXCEPT?

a. Retinoschisis – Hyperopia
b. Goldman Favre – Myopia
c. Stickler – Myopia
d. Wagner – Myopia

17. Which is likely to be the imaging modality of choice in the following patient?

A 17-year-old boy who was in a motor vehicle accident hit his face on the dashboard. He has blurring of vision and a relative afferent pupillary defect.

a. MRI with STIR sequence
b. Contrasted CT scan of the brain and orbit
c. Plain CT scan of the brain and orbit
d. MRI with a FLAIR sequence

18. The following are features of the LogMAR chart EXCEPT?

a. Linear progression in letter size
b. Equal number of letters per line
c. Regular spacing between lines and letters
d. Ensures letter shape is the sole determinant of difficulty of each line

19. Which of the following statements is MOST likely to be false regarding ophthalmic ultrasonography?

a. 5 spikes are seen on a contact A scan
b. An immersion A scan requires the patient to be supine
a. A B scan is unable to reliably distinguish between vitritis and a vitreous haemorrhage
d. A silicone oil filled eye will result in a artificially short axial length

20. Of the following investigative results, which is LEAST likely to be true in a patient with thyroid eye disease?

a. Thyroid stimulating hormone (TSH) receptor antibodies are more likely to be present as compared to thyroglobulin antibodies
b. Tendon involvement on orbital imaging is not seen
c. On fields of binocular single vision, diplopia is most commonly seen during adduction of the involved eye in unilateral , burnt out thyroid eye disease
d. Relative lymphocytosis is often present

21. The following statements regarding the mode of spread by choroidal melanoma are TRUE except?

a. Haematogenous metastasis is most commonly to the liver
b. Optic nerve extension is common
c. Orbital metastasis may occur by local extension through transcleral emissary channels
d. Lymphatic spread is not possible

22. Which of the following statements is TRUE regarding cholinergic drugs?

a. Acetylcholine is stable in the anterior chamber
b. Intracameral acetylcholine has a less rapid onset than intracameral carbachol
c. Pilocarpine 1% is used to diagnose Adie's tonic pupil
d. May cause reduced acuity in scotopic conditions

23. Which of the following drugs is MOST likely to be responsible for the scenario below?

A 73-year-old lady presents 2 weeks after cataract surgery with a painful red eye. Her vision is normal. She has been on topical gentamicin/betamethasone combination drops which she could not tolerate.
She has since stopped the drops and has been self-prescribing topical levofloxacin which she bought over the counter, together with non preserved artificial tears.
On examination, she has generalized hyperaemia, pseudomembrane and confluent punctate corneal erosions.

a. gentamicin
b. betamethasone
c. levofloxacin
d. artificial tears

24. The following statements regarding the side effects of acetazolamide are TRUE except?

a. May cause Steven Johnson syndrome
b. Patients susceptible to aplastic anemia may be detected with pre-treatment blood counts
c. Known to cause limb deformity in animal tests
d. May cause depression

25. The following are typical causes of chronic post-operative endophthalmitis EXCEPT?

a. Propionibacterium acnes
b. Corynebacterium spp.
c. Fungi
d. Staphylococcus aureus

26. Regarding Stevens Johnson Syndrome (SJS), which of the following statements is MOST likely to be false?

a. Is characterized by a Type 2 hypersensitivity reaction with antibody formation against epithelial basement membrane antigens
b. Human Immunodeficiency Virus (HIV) infected individuals are more susceptible
c. Typically involves the skin
d. May be triggered by acetazolamide

27. Which of the following features is MOST suggestive of chlamydial inclusion conjunctivitis in an adult?

a. Bulbar conjunctival follicles
b. Serous discharge
c. Pre-auricular lymphadenopathy
d. Punctate epithelial erosions

28. Which of the following statements regarding neovascular glaucoma (NVG) is MOST likely to be false?

a. May occur in chronic retinal detachments
b. Occurs in up to 50% of patients with ischaemic central retinal vein occlusion (CRVO)
c. Topical pilocarpine is the drug of choice in early NVG
d. Intravitreal bevacizumab has been shown to be efficacious in controlling intraocular pressure in early NVG

29. Which of the following disorders is the MOST likely diagnosis for the following scenario?

A 29-year-old stockbroker presents with sudden onset blurring of vision in his left eye. Clinically, his right eye is normal, and his left eye shows a small macular elevation with no sub-retinal blood. His vision is OD 6/6 OS 6/9. Refraction is OD: plano OS: +0.75.

a. Age Related Macular Degeneration (AMD)
b. Myopic Maculopathy
c. Central Serious Chorioretinopathy (CSCR)
d. Idiopathic Polypoidal Choroidal Vasculopathy (IPCV)

30. Of the following disorders, which is the MOST likely diagnosis for the scenario below?

A 6-year-old girl with deafness presents with iris heterochromia and pigmentary retinopathy. She has a white tuft of hair.

a. Hurler-Scheie
b. Batten Disease
c. Alport Syndrome
d. Waardenburg

31. Which of the following is LEAST likely to be false about Adult Foveolar Vitelliform Dystrophy?

a. Is autosomal recessively inherited
b. No phenotypic variability
c. Onset between ages of 30-50 years
d. May affect peripheral vision

32. According to 2012 Royal College guidelines for the management of diabetic macular oedema, which of the following statements is FALSE?

a. Grid laser remains the mainstay of treatment
b. Grid laser alone does not usually result in significant improvement in visual acuity
c. Grid laser photocoagulation should be considered first line treatment in patients with centre-involving macular oedema and a visual acuity of better than 78 letters
d. Intravitreal ranibizumab is indicated as first line treatment in patients with centre-involving macular oedema and a visual acuity of worse than 78 letters

33. In which of the following scenarios may a doctor breach a patient's confidentiality without their express permission?

a. The police are seeking information about an accident 6 months ago in which a patient was involved
b. A lawyer contacts you regarding information about one of your patients without their knowledge
c. The police are seeking information regarding one of your patients who was last night, involved in a struggle with an armed criminal whose whereabouts are as yet, unknown
d. A patient's wife calls you regarding information about her husband's diabetic retinopathy

34. Regarding the following data set ' 2,2,2,2,3,3,4,6,8,10', which of the following statements is FALSE?

a. Demonstrates a positive skew
b. The mean is 4.2
c. The mode is 2
d. Demonstrates a negative skew

35. The following statements regarding myotonic dystrophy are TRUE except?

a. Is autosomal dominantly inherited
b. Gastrointestinal symptoms occur in at least 50% of patients
c. Lens changes include scintillating, particle – like anterior and posterior subcapsular cataract
d. 70% of associated mortality is due to aspiration pneumonia and sepsis

36. The following are American College of Rheumatology (ACR) criteria for diagnosing Giant Cell Arteritis (GCA) EXCEPT?

a. Age ≥ 50 years at disease onset
b. Elevated Erythrocyte Sedimentation Rate (ESR) >50mm/h
c. Female Sex
d. Temporal artery tenderness

37. Which of the following statements regarding neuromyelitis optica (NMO) is FALSE?

a. Radiologic findings include paucity of cerebral demyelinating lesions
b. More than 3 spinal cord segments are often involved
c. Cerebrospinal fluid analysis (CSF) typically reveals oligoclonal bands
d. Typically presents with recurrent spinal cord myelitis

38. The following statements regarding the presentation of alcohol tobacco amblyopia are likely to be TRUE except?

a. The papillomacular bundle appears to be particularly susceptible to under nutrition in association with tobacco or alcohol abuse
b. The classic field defects are arcuate scotomas
c. Pain is usually absent
d. Colour vision defects are common

39. Which of the following is the MOST likely diagnosis for the given scenario?

A 65-year-old gentleman with a history of laparotomy and total gastrectomy presents with bilateral dry eyes. He also has significant bilateral upper limb weakness.

a. Lambert Eaton Myasthenic Syndrome (LEMS)
b. Botulism
c. Myasthenia Gravis (MG)
d. Hypothyroidism

40. Regarding sebaceous gland carcinoma, please choose the INCORRECT statement.

a. Comprise 1% of eyelid malignancies
b. Occurs more commonly in men than in women
c. Tend to involve the upper lid
d. Clinical appearance is variable and may mimic benign conditions such as chronic blepharitis

41. What is the most common location for eyelid pseudo-colobomas in Treacher Collins Syndrome?

a. Medial Upper Lid
b. Lateral Upper Lid
c. Medial Lower Lid
d. Lateral Lower Lid

42. For the following scenario, which lid reconstructive procedure is MOST likely to be efficacious?

A 65-year-old gentleman presents with a suspicious lesion on his right upper eye lid. Moh's surgery and conjunctival map biopsies are performed, and the gentleman is left with a full thickness defect of the upper lid of 80% horizontal length. The vision in his both eyes is 6/9 uncorrected.

a. Hughes Tarsoconjunctival Flap and Skin Graft
b. Cutler Beard bridge flap
c. Mustarde's Cheek Rotation Flap
d. Tenzel's Semicircular Flap

43. The following conditions may be responsible for hypotony in a traumatised eye EXCEPT?

a. Ciliary body shock
b. Hyphaema
c. Ciliary body detachment
d. Inflammation

44. How often does senile retinoschisis involve the infero-temporal quadrant?

a. 50%
b. 60%
c. 70%
d. 80%

45. In which of the following scenarios is intravitreal ranibizumab MOST appropriate?

a. Macular oedema of an area 2 mm in diameter encroaching within 500μm of the fovea with acuity >78 letters
b. Macular oedema involving the centre of the fovea with visual acuity >78 letters
c. Macular oedema involving the centre of the fovea with visual acuity 50 letters
d. Macular oedema involving the centre of the fovea with visual acuity <24 letters

46. Which of the following is NOT an indication for adjustable sutures in strabismus surgery?

a. Correction of esotropia in burnt out thyroid eye disease
b. Jensen's procedure in correction of a 6th nerve palsy
c. Bilateral medial rectus recession in congenital esotropia
d. Hypotropia correction in patients with history of blow out fractures of the orbital floor

47. Which of the following would be the location of choice to harvest a graft for the following scenario?

A 70-year-old lady presents with a suspicious lesion on her left upper eyelid. The lesion is removed by splitting the lid at the grey line, followed by wide local excision, necessitating an anterior lamellar graft. She has mild lagophthalmos of the contralateral upper eyelid due to scarring from a previous thermal injury.

a. Preauricular skin
b. Split skin graft
c. Postauricular skin
d. Abdominal skin

48. A 6-year-old girl who has been diagnosed with juvenile idiopathic arthritis (JIA) has been followed up in your clinic every 4 months for the past 2 years.
The paediatric rheumatologist has discontinued her methotrexate the week before.
She is asymptomatic, without any ocular inflammation. What is the most appropriate course of action?

a. Continue follow ups 4 monthly up to 12 years of age
b. Continue follow-ups at 2 monthly intervals for 6 months
c. Discontinue follow-ups
d. Follow-up at a week

49. The following are parametric tests EXCEPT?

a. One Way ANOVA
b. Paired T Test
c. Wilcoxon Signed Ranks Test
d. Independent T Test

50. Regarding adrenal crisis, which of the following statements is MOST likely to be incorrect?

a. May be seen in patients with abrupt withdrawal of long term steroid therapy
b. Typically manifests with acute hypovolemic shock
c. Administration of physiologic doses of steroids and aggressive fluid management is the established therapy
d. May be complicated with hypokalaemia

51. Which of the following statements regarding Irvine Gass Syndrome is MOST likely to be false?

a. Risk of occurrence is higher with anterior chamber or sulcus intraocular lenses as compared to in-the-bag implantation
b. The vast majority of cases resolve spontaneously within a year
c. Diabetes mellitus is a risk factor
d. Fluorescein angiography typically reveals well-defined 'ink blot' hyperfluorescence

52. Regarding Thygeson's Superficial Punctate Keratitis (TSPK), please choose the CORRECT statement.

a. Typically unilateral
b. There is often no conjunctival inflammation
c. The stroma is often involved
d. Intraepithelial whitish lesions that stain with fluorescein are typical

53. Which of the following statements regarding optic nerve hypoplasia (ONH) is MOST likely to be true?

a. Maternal diabetes mellitus is not a risk factor
b. De Morsier syndrome is the midline brain defect most commonly associated with ONH
c. Is not associated with midline bony defects
d. Growth retardation is not an associated condition

54. The following are risk factors for Herpes Zoster Ophthalmicus (HZO) EXCEPT?

a. Renal transplant patients
b. Afro-Caribbean race
c. Infection with Human Immunodeficiency Virus (HIV)
d. Patient age over 60

55. The following statements regarding orbital metastases are TRUE except

a. Less common than uveal metastases
b. Comprise less than 10% of orbital neoplastic lesions
c. Thyroid carcinoma is the most common primary tumour
d. Diplopia is often the presenting feature

56. Which of the following statements regarding isolated, unilateral lesions of the lateral geniculate body (LGB) is MOST likely to be true?

a. Causes decreased visual acuity
b. Presence of a contralateral relative afferent papillary defect (RAPD)
c. Absence of optic disc pallor
d. Contralateral limb weakness may be present

57. Which of the following features of Tolosa Hunt Syndrome is MOST likely to be false?

a. Is characterised by severe retrobulbar pain
b. Spontaneously resolves without treatment
c. Attacks recur within months to years
d. MRI findings are pathognomonic

58. Which of the following pairs of inherited connective tissue disorders and affected protein is INCORRECT?

a. Pseudoxanthoma Elasticum – Elastin
b. Osteogenesis Imperfecta – Type 1 Collagen
c. Stickler's Syndrome – Type 9 Collagen
d. Marfan's Syndrome – Fibrillin

59. Which of the following statements regarding intermittent exotropia is LEAST likely to be true?

a. Commonest cause of exotropia
b. Presents most commonly from 8 years onwards
c. Patients often have excellent stereopsis
d. Surgery is usually indicated to restore binocular single vision

60. A 6–year-old boy who wears thick concave glasses is referred from his optician for reduced spectacle acuity. On examination, his vision is OD: 6/36 OS: 6/24. He has nystagmus and his fundal examination is normal. His D15 testing consistently reveals errors along the protan and deutan axes.

Which of the following conditions is MOST likely to be responsible?

a. Blue Cone Monochromatism
b. Oguchi's Disease
c. Rod Monochromatism
d. Alstrom Syndrome

61. Which of the following statements regarding the pathophysiology and clinical presentation of alphabet pattern deviations is MOST likely to be false?

a. The commonest cause of V pattern strabismus is inferior oblique (IO) over action
b. V patterns may be acquired after head trauma
c. The angle of deviation in an A pattern diverges by 15 prism dioptres (PD) or more from up gaze to down gaze
d. Inferior oblique myotomy is a treatment option for V pattern esodeviations secondary to IO over action

62. What is the proportion of patients with Stage 1 macular holes that spontaneously resolve?

a. 30%
b. 40%
c. 50%
d. 60%

63. Regarding viscoelastics, which of the following statements regarding their behaviour is MOST likely to be false?

a. Molecular weight is an important determinant of the ability of the viscoelastic to remain in the eye during irrigation
b. Sodium hyaluronate coats intraocular surfaces well
c. Peak pressure rise from retained viscoelastic occurs after 24 hours post-cataract surgery
d. Viscoelastic aspiration after lens implantation improves trabecular outflow

64. Regarding Posterior Polymorphous Dystrophy (PPMD), which of the following statements is INCORRECT?

a. Is autosomal dominantly (AD) inherited
b. Often shows rapid progression from asymptomatic to symptomatic disease
c. Characteristic findings include posterior corneal surface vesicles
d. May be confused with iridocorneal endothelial syndrome (ICE)

65. The following are risk factors for pseudoexfoliation syndrome EXCEPT?

a. Age
b. Southern Latitudes
c. Women
d. Higher Altitudes

66. Which of the following disorders is MOST likely to present as a poorly defined mass on orbital imaging?

a. Cavernous Haemangioma
b. Dacryops
c. Orbital Metastases
d. Lymphangioma

67. Which of the following features is LEAST likely to be suggestive of malignant change within a conjunctival nevus?

a. Change in colour
b. Presence of intralesional cysts
c. Fornix location
d. Scleral fixation

68. Regarding glaucoma in pseudoexfoliation syndrome, which of the following statements is MOST likely to be true?

a. Glaucoma tends to present unilaterally
b. Has an angle closure component in over 50% of patients
c. Has a more benign course as compared to primary open angle glaucoma
d. Selective Laser Trabeculoplasty (SLT) is not a treatment option

69. Which of the following corneal dystrophies is visually significant?

a. Central Cloudy Dystrophy
b. Meesmans Dystrophy
c. Posterior Amorphous Corneal Dystrophy
d. Congenital Hereditary Stromal Dystrophy

70. Which of the following statements regarding interstitial keratitis (IK) is MOST likely to be false?

a. Is most commonly associated with *Treponema pallidum*
b. May present together with hearing loss in association with rheumatoid arthritis
c. Onset often involves both eyes simultaneously
d. There is stromal involvement without significant involvement of the epithelium or endothelium

71. Angioid streaks are associated with defects at which layer of the retina?

a. Bruch's Membrane
b. Internal Limiting Membrane
c. External Limiting Membrane
d. Choroid

72. Which of the following statements regarding sickle cell retinopathy is MOST likely to be incorrect?

a. Sickle Cell disease (HbSS) is more common than Sickle Cell Thalassaemia (HbSThal)
b. Patients with Sickle Cell Haemoglobin C (HbSC) disease are more likely to have retinal manifestations as compared to patients with sickle cell (HbSS) disease
c. Auto infarction of peripheral retinal vessels may occur in as many as 60% of eyes
d. Direct treatment of nutrient arterioles and draining venules is contraindicated

73. Regarding Keratoconus, which of the following statements is MOST likely to be false?

a. There is evidence of familial inheritance in up to 15% of cases
b. Ocular associations include Leber's Hereditary Optic Neuropathy
c. Wearing of both rigid and soft contact lenses are risk factors
d. Genetic evidence of abnormal antioxidative-stress related mechanisms has been reported

74. According to the results of the following investigations, what is the MOST likely diagnosis?

A 40-year-old lady presents with unilateral blurring of vision. She has a well-demarcated, orange fundal lesion next to the optic disc in her right eye. There is macular oedema. Ultrasound reveals a highly reflective anterior border with high internal reflectivity. There is no orbital shadowing. Fluorescein angiography reveals early hyperfluorescence of the lesion that increases throughout the angiogram with diffuse, late leakage.

a. Choroidal Melanoma
b. Choroidal Haemangioma
c. Melanocytoma
d. Uveal Lymphoma

75. Regarding Bell's palsy, which of the following features is MOST likely to be present?

a. Sparing of the frontalis muscle
b. Normal eyelid closure
c. Lacrimal pump failure causing epiphora
d. Normal taste

76. According to the following scenario, what would be the MOST likely site of the offending lesion?

A 30-year-old lady presents with anisocoria that is worse in the dark. She does not have any history of facial dryness. Apraclonidine 0.5% test is positive. 2 days later, a hydroxyamphetamine 1% test does not result in pupillary dilation.

a. Occlusion of the superior cerebellar artery
b. Dissection of the thoracic aorta
c. Pontine infiltration by sarcoidosis
d. Intracavernous carotid artery aneurysm

77. The axial length of the neonatal eye is

a. 15.5 mm
b. 16.0 mm
c. 16.5 mm
d. 17.0 mm

78. The following are causes of autofluorescence EXCEPT?

a. Drusen
b. Astrocytomas
c. Lipofuscin
d. Choroidal naevi

79. Which of the following statements is LEAST likely to be correct regarding the features of HLA B27 related uveitis?

a. Typically presents unilaterally
b. Is associated with an increased number of recurrences as compared to idiopathic acute anterior uveitis (AAU)
c. Tends to be associated with more severe inflammation as compared to idiopathic AAU
d. Occurs with similar frequency in both males and females

80. Regarding electrodiagnostic testing in Myasthenia Gravis (MG), which of the following statements is MOST likely to be false?

a. Single fiber electromyography (SFEMG) is more sensitive than repetitive nerve stimulation (RNS)
b. The most common stimulation rate in RNS is 10Hz
c. Pyridostigmine may mask RNS abnormalities
d. SFEMG abnormalities are not masked by pyridostigmine

81. What is likely to be the MOST helpful investigation for the following scenario?

A 27-year-old lady with a BMI of 21 presents with bilateral, intermittent visual loss. She is on oral contraceptives. She has frontal headaches that are worse when bending over. She has bilateral optic disc swelling. An urgent non-contrasted computerized tomographic (CT) scan shows no abnormality with normal sized ventricles. Lumbar puncture opening pressure is 28 cmH2O.

a. Magnetic Resonance Angiography (MRA) of the Circle of Willis
b. Magnetic Resonance Venography (MRV) of the venous sinuses
c. Contrasted Computed Tomographic Scan of the Brain
d. Computed Tomographic Angiogram of the Circle of Willis

82. Which is likely to be the MOST helpful imaging method of choice for the following patient?

A 61-year-old gentleman presents with acute onset proptosis of his right eye, with periorbital swelling and redness. He has profound loss of vision, a relative afferent pupil defect and ocular motility is limited in all directions. He is febrile and drowsy. Within 12 hours, he develops lid swelling of his left eye.

a. Contrasted Computed Tomographic (CT) scan of the orbits
b. Magnetic Resonance Venography (MRV)
c. Gadolinium Enhanced Magnetic Resonance Imaging (MRI) scan of the orbits
d. Non-contrasted CT scan of the orbits

83. According to Royal College of Ophthalmology referral guidelines for Ocular Tumors, which of the following features is NOT one that should prompt a referral?

a. Conjunctival melanocytic tumor of more than 3.0 mm with clear cysts
b. Choroidal tumor with a thickness of more than 2.0 mm
c. Choroidal tumor with documented growth
d. Conjunctival melanocytic tumor with nodularity

84. Regarding suturing of corneal laceration wounds, please choose the CORRECT statement.

a. Suturing of a limbus-to-limbus corneal laceration should begin at the centre of the wound, working laterally towards the limbus.
b. A 6-0 polyglactin suture swaged to a 3/8 spatulated needle is the suture of choice
c. The needle should enter the corneal tissue at a 90 degree angle to the corneal surface
d. The sutures should be shortest at the corneal limbus and longest in the centre.

85. A 40-year-old lady with active thyroid eye disease is admitted for blurring of vision in her right eye. On examination, her right eye is proptosed with an intraocular pressure of 24 mmHg and a relative afferent pupillary defect. Her right optic disc is swollen. Examination of her left eye is normal.

What is the MOST appropriate treatment option?

a. Oral prednisolone 1mg/kg OD
b. Pulsed Intravenous Methylprednisolone
c. Orbital Decompression
d. Retrobulbar Steroids

86. Regarding the following scenario, which of the following nystagmoid movements is MOST likely to be seen?

A 6-year-old boy with Arnold Chiari malformation and hydrocephalus presents with involuntary ocular movements. On examination, the movements are conjugate, with a slow upward pursuit movement and fast corrective downward saccade.

a. Upbeat nystagmus
b. Gaze evoked nystagmus
c. Downbeat nystagmus
d. Acquired vertical nystagmus

87. The following are causes of hypofluorescence on fluorescein angiography EXCEPT?

a. Pre-macular haemorrhage
b. Macular ischemia
c. Congenital Hypertrophy of the Retinal Pigment Epithelium
d. Neovascular Age Related Macular Degeneration (AMD)

88. Which of the following investigations is likely to be the imaging method of choice for the following patient?

A 29-year-old lady presents with sudden onset painful blurring of vision in her right eye. Her vision has been deteriorating over the past week. There is optic disc swelling and a relative afferent pupillary defect.

a. Contrasted computed tomographic (CT) scan of the brain and orbits
b. Plain CT scan of the brain and orbits
c. Non-contrasted magnetic resonance imaging (MRI) of the brain
d. MRI with a FLAIR sequence.

89. Pertaining to the HIV virus, which statement is MOST likely to be false?

a. Infection with HIV-2 tends to progress to AIDS faster than infection with HIV-1
b. It is a single stranded RNA virus of the Retroviridae family
c. Co-infection with Human Papillomavirus (HPV) is common
d. HIV-2 infection is more common worldwide

90. Of the following investigations in Hashimoto's Disease, which is LEAST likely to occur?

a. Low Thyroid Stimulating Hormone (TSH) levels
b. Low free T4 levels
c. Absence of TSH receptor antibodies
d. Presence of Thyroid Peroxidase (TP) antibodies

Paper 5 Answers and Discussion

1. Answer: c

Discussion: Sturge Weber Syndrome occurs sporadically, and thus, best fits scenario c.
Scenario a, with an occurrence amongst the siblings of 50% with an affected parent, likely demonstrates an autosomal dominant inheritance pattern.
Scenario b, with unaffected parents and an occurrence amongst siblings of 25%, likely demonstrates an autosomal recessive inheritance pattern.
Scenario d, with unaffected parents and 3 boys plus a maternal uncle who are affected demonstrates an X linked recessive inheritance pattern.

2. Answer: c

Discussion: A review in 2003 of epidemiologic studies concerning burn wound infections revealed that the most important isolates were *Pseudomonas aeruginosa*. Antibiotic therapy should include an extended spectrum penicillin plus an aminoglycoside, subject to change according to cultures and sensitivities.

Source:
1. Mayhall CG. The Epidemiology of Burn Wound Infections: Then and Now. *Clin Infect Dis.* (2003) 37 *(4): 543-550.*

3. Answer: c

Discussion: All points within Panum's area fall on non-corresponding retinal elements. However, fusion allows binocular single vision with stereopsis. Points lying outside Panum's area are seen as separate points.

Source:
1. 2009-2010. *Pediatric Ophthalmology and Strabismus.* American Academy of Ophthalmology

4. Answer: a

Discussion: Touton giant cells are seen in xanthogranuloma. Langhan giant cells are seen in sarcoidosis, tuberculosis and Giant Cell Arteritis.

5. Answer: c

Discussion: Scenario 'a' describes an autosomal dominant trait with one parent affected and 50% of children affected.
Scenario 'b' describes a sporadic trait with no family history.
Scenario 'c' describes a mitochondrial-inherited trait with all siblings affected, as well as a positive history involving their mother and her siblings.
Scenario 'd' may describe a mitochondrial-inherited trait. However, in this scenario, those affected are men and thus is more likely to describe an x linked recessive trait.

Source:
1. Yu PWM, Chinnery PF. Leber Hereditary Optic Neuropathy. *Gene Reviews.* October 26, 2000

6. Answer: d

Discussion: As x-linked disorders with ocular associations are generally less common as compared to conditions with a Mendelian inheritance, it makes sense to memorise these.

(mnemonic = I **H**ad to dump my e**X** as she/he loved to **LOAF**) **H**unter, **X**linked, **L**owe's, **O**cular Albinism, **A**lport's, **F**abry's)
Homocystinuria is autosomal recessively inherited.

7. Answer: c

Discussion: In homocystinuria, the direction of subluxation is often inferonasal due to brittle zonules that rupture, allowing the lens to drop following the pull of gravity.

8. Answer: b

Discussion: Adenoviral conjunctivitis is the commonest cause of acute conjunctivitis in school going children, and may present in one of two main clinical pictures.
Pharyngoconjunctival fever, typically affecting young children, presents with systemic upset and rarely, with subepithelial infiltrates.
Epidemic keratoconjunctivitis, which typically affects adults, rarely presents with systemic upset, and subepithelial infiltrates are common.
Both typically present with a follicular conjunctivitis and tender pre-auricular lymphadenopathy.

While acyclovir is not effective, topical ganciclovir has shown some promise.

Source:
1. Colin J. Ganciclovir ophthalmic gel, 0.15%: a valuable tool for treating ocular herpes *Clin Ophthalmol*. 2007 December; 1(4): 441–453.

9. Answer: c

Discussion: Sturge-Weber Syndrome carries a glaucoma risk of 50%. Patients at increased risk include patients whose port wine stain involves the distribution of the ophthalmic division of the trigeminal nerve.

Source:
1. Sharan S, Swamy B, Taranath DA, Jamieson R, Yu T, Wargon O, Grigg JR. Port-wine vascular malformations and glaucoma risk in Sturge- Weber syndrome. *J AAPOS*. 2009 Aug;13(4):374-8

10. Answer: d

Discussion: The features suggestive of an ischaemic CRVO are a visual acuity of <6/60, an RAPD, widespread deep intraretinal haemorrhages, >10 cotton wool spots, an electronegative ERG and more than 10 disc diameters of capillary fall out on fluorescein angiography.

11. Answer: c

Discussion: NA-AION is the most common cause of acute optic neuropathy in patients over 50. The pathology involves acute infarction of the optic nerve head. Risk factors include diabetes and hypertension, and because patients classically present with painless visual loss upon waking up, nocturnal hypotension is hypothesized to play a major role. Classic findings include sectoral oedema of the superior half of the disc with corresponding inferior altitudinal field defects.

12. Answer: c

Discussion: The prognosis of BCC is generally excellent, with cure rates approaching 95%. There is a 5% overall recurrence rate, which depends on type of treatment and histologic subtype. Incidence of metastasis is in the region of 0.1% with lungs and bones being the most common sites. Over 5 years, aside from the 5% recurrence rate, there is a 35% risk of developing BCCs in other sites.

Source:
1. McLoone NM, Tolland J, Walsh M, et al. Follow up of basal cell carcinomas: an audit of current practice. *J Eur Acad Dermatol Venereol*. Jul 2006;20(6):698-701

13. Answer: c

Discussion: The classic triad of congenital toxoplasma infection is chorioretinitis, hydrocephalus and intracranial calcifications. The chorioretinitis, which occurs in 79% of infected neonates, involves the macula in at least 50% of patients.

Source:
1. Mets MB, Chhabra MS. Eye Manifestations of Intrauterine Infections and Their Impact on Childhood Blindness. *Surv Ophthal*. Vol 53(2) March – April 2008

14. Answer: a

Discussion: Micro esotropias are small angled deviations, usually measuring less than 10 prism diopters. They are characterised by a small central fixation scotoma, detectable by the 4-prism dioptre test. However, there is often evidence of para foveal fixation that allows some degree of binocular function.

Source:
1. MacEwen C, Gregson R. 2003. *Manual of Strabismus Surgery*. London. Elsevier Limited

15. Answer: a

Discussion: The exact pathogenesis of traumatic macular holes is as yet undetermined. Traumatic macular holes occur within hours to weeks of trauma, and are typically associated with ocular contusions. They have a higher spontaneous closure rate as compared to idiopathic holes, and in patients whom surgical closure is indicated, vitrectomy, ILM peel and gas tamponades are highly effective.

Source:
1. Kuhn F, Morris R, Mester V, Witherspoon C (2000) Internal limiting membrane removal for traumatic macular holes. *Ophthalmol Surg Laser* 31: 308–315
2. Yamada H, Sakai A, Yamada E, Nishimura T, Matsumura M (2002) Spontaneous closure of traumatic macular hole. *Am J Ophthalmol* 134: 340–347

16. Answer: b.

Discussion: Both retinoschisis and Goldman Favre are associated with hyperopia, whereas Stickler and Wagner are associated with myopia.

Source:
1. Tasman W, Jaeger EA. 2013. *Duane's Ophthalmology*. Lippincott Williams & Wilkins.

17. Answer: c

Discussion: This is likely to be traumatic optic neuropathy. Imaging method of choice would be a plain CT scan with thin 1.5 mm slices which should detect retrobulbar haemorrhage, fractures of the sphenoid or a nerve sheath haematoma which are common causes of traumatic optic neuropathy. Contrast is non-essential.

18. Answer: d

Discussion: The way the LogMAR chart was designed allows avoidance of particular disadvantages of the Snellen chart, for example, the fact that certain optotypes were easier to recognize than others. LogMAR utilizes optotypes of similar shape, ensuring that letter size alone was the determinant of line difficulty.

19. Answer: d

Discussion: Immersion A scans require a water bath and hence, require the patient to be supine.
A silicone oil filled eye will lead to an artificially long eye.

20. Answer: c

Discussion: TSH antibodies are seen more frequently than thyroglobulin antibodies in patients with thyroid eye disease.
Orbital imaging typically shows multiple recti enlargement with tendon sparing. Diplopia is commonest in elevation and abduction of the involved eye in unilateral disease because the inferior and medial recti are most commonly involved, leading to shortening of muscles and tethering of the globe. Blood investigations often reveal a normocytic anaemia, and low total white count with relative lymphocytosis.

21. Answer: b

Discussion: Choroidal melanomas may spread in 3 ways; haematogenous (which is most common, and commonly to the liver), transclerally into the orbit via emissary channels, and rarely, via the optic nerve. (which occurs only with peripapillary tumors) Lymphatic spread is not possible as the eye lacks lymphatic channels.

22. Answer: d

Discussion: Intracameral carbachol during cataract surgery induces miosis which helps to avoid optic capture and iris incarceration in wounds. Acetylcholine, which has a more rapid onset, is broken down rapidly and therefore has a very short duration of action.

Pilocarpine 0.1% aids diagnosis of Adie's pupil. Pilocarpine 2-4% may be used as short-term prophylaxis against attacks of acute angle closure, but is not recommended as long-term replacement for an iridotomy. Amongst the side effects of cholinergic drugs is reduced acuity at night, which may be disabling.

23. Answer: a

Discussion: Of the above, the likeliest culprit would be topical gentamicin. The aminoglycosides are quite toxic to the corneal epithelium, and neomycin in particular, may cause an oculocutaneous allergic reaction.

Source:
1. Abott, G.L, Morrow, R.L. Conjunctivitis. *Am Fam Physician*. 1998 Feb 15;57(4):735-746.

24. Answer: b

Discussion: The side effect profile of acetazolamide is significant, with up to a half of patients unable to tolerate them. Potentially life-threatening side effects include thrombocytopenia, Steven Johnson Syndrome, aplastic anemia (susceptibility cannot be ascertained from blood counts) and metabolic acidosis. Teratogenicity is well documented in rodents.

25. Answer: d

Discussion: *Staphylococcus aureus* is seen typically in acute post-operative endophthalmitis.

Source:
1. Hanscom TA. Postoperative Endophthalmitis. *Clinical Infectious Diseases* 2004; 38:542–6

26. Answer: a

Discussion: SJS is characterized by blistering of the skin and mucus membranes caused by a Type 3 hypersensitivity reaction. SJS may be triggered by infection (commonest cause in pediatric cases), drugs, malignancies, and may be idiopathic in up to 50% of cases. Prominent drug causes include antivconvulsants, allopurinol, non-steroidal anti inflammatory drugs (NSAIDS) and sulfa based drugs like acetazolamide.

Source:
□□□Rotunda A, Hirsch RJ, Scheinfeld N, Weinberg JM. Severe cutaneous reactions associated with the use of human immunodeficiency virus medications. *Acta Derm Venereol*. 2003;83(1):1-9.
□□□Gruchalla RS. Drug allergy. *J Allergy Clin Immunol*. Feb 2003;111(2 Suppl):S548-59

27. Answer: a

Discussion: Features suggestive of chlamydial inclusion conjunctivitis include pre-auricular lymphadenopathy, chemosis, inferior tarsal follicles, bulbar conjunctival follicles, mucoid discharge and punctate erosions.
While the other features may be seen in various conditions, bulbar follicles are a distinctive feature of chlamydial conjunctivitis.

Source:
1. American Academy of Ophthalmology Corneal/External Disease Panel. Preferred Practice Pattern: Conjunctivitis. San Francisco, Ca: AAO; 2003.

28. Answer: c

Discussion: NVG is caused by anterior segment neovascularisation triggered by retinal ischemia. The commonest causes of NVG include proliferative diabetic retinopathy and ischemic central retinal vein occlusions. Treatment options for early disease include topical atropine and antiglaucoma agents. Pilocarpine is contraindicated as it tends to cause inflammation. Intravitreal bevacizumab is effective in stabilizing neovascularisation and controlling intraocular pressure in early disease. Surgery is indicated in late disease as medical therapy is often not effective.

Source:
1. Martinez-Carpio PA, Bonafonte-Marquez E, Heredia-Garcia CD, Bonafonte-Royo S. [Efficacy and safety of intravitreal injection of bevacizumab in the treatment of neovascular glaucoma: systematic review]. *Arch Soc Esp Oftalmol*. Oct 2008;83(10):579-88.
2. Wakabayashi T, Oshima Y, Sakaguchi H, Ikuno Y, Miki A, Gomi F, et al. Intravitreal bevacizumab to treat iris neovascularization and neovascular glaucoma secondary to ischemic retinal diseases in 41 consecutive cases. *Ophthalmology*. Sep 2008;115(9):1571-80

29. Answer: c

Discussion: This scenario is likely to point to central serous chorioretinopathy, given the age and high stress occupation of the patient. The mild hypermetropia due to macular elevation is also typical of CSCR. Neovascular AMD and IPCV both typically present with sub retinal haemorrhage, although IPCV tends to be unilateral, occurring in younger, Asian patients, while AMD tends to be bilateral and occurs in older patients.

Source:
1. Uyama M. et al. Idiopathic Polypoidal Choroidal Vasculopathy in Japanese patients. *Arch Ophthalmol. 1999;117(8):1035-1042.*

30. Answer: d

Discussion: Waardenburg's is the likeliest answer. It is an autosomal dominant disorder characterized by failure of melanocyte differentiation. This affects hearing, as cochlear function is dependent on melanocyte differentiation. Pigmentation of the iris as well as retina is thus also abnormal. Features include white forelock, upturned nose, short upper lid, telecanthus, iris heterochromia and pigmentary retinopathy.

Source:
1. emedicine.medscape.com

31. Answer: c

Discussion: Adult Foveolar Vitelliform Dystrophy is caused by mutations in the *PRPH2* or *BEST2* genes. It is characterised by autosomal dominant inheritance with phenotypic variability. Onset is between 30-50 years of age with mild visual loss and yellow foveal deposits at the level of the retinal pigment epithelium. It does not affect the peripheral vision or cause nyctalopia.

Source:
1. Epstein GA, Rabb MF.Adult vitelliform macular degeneration: diagnosis and natural history. *Br J Ophthalmol*, 1980,64, 733-740

32. Answer: d

Discussion: Intravitreal ranbizumab should be considered as first line treatment, with or without deferred grid laser photocoagulation, in patients with centre involving macular oedema, a central foveal thickness of $250\mu m$ or more, and a visual acuity of worse than 78 letters.

Source:
1. Royal College of Ophthalmologists Guidelines for Diabetic Retinopathy 2012

33. Answer: c

Discussion: Scenarios where confidentiality may be breached without a patient's express permission would include situations where the doctor believes that without volunteering said information, the patient themselves or other people may be at risk of harm.

Source:
1. General Medical Council Guidelines on Confidentiality October 2009

34. Answer: d

Discussion: The data set demonstrates a positive skew, with a tail on the right.

35. Answer: d

Discussion: 70% of myotonic dystrophy associated mortality is due to cardiac arrhythmias.

Source:
1. Bradley WG. 2004. *Neurology in Clinical Practice: Principles of diagnosis and management.* 4th Ed. Philadelphia. Elsevier Inc.

36. Answer: c

Discussion: ACR criteria for diagnosing GCA are 3 of the following 5: age ≥ 50 at disease onset, ESR>50mm/h, new onset localized headache, temporal artery tenderness and a positive arterial biopsy. The presence of 3 features carries a sensitivity of 93.5% and a specificity of 91.2%.

Source:
1. Hunder GG, Bloch DA, Michel BA, Stevens MB, Arend WP, Calabrese LH, et al. The American College of Rheumatology 1990 criteria for the classification of giant cell arteritis. *Arthritis Rheum* 1990;33:1122‑‑‑8

37. Answer: c

Discussion: NMO is a severe, demyelinating disease that selectively involves the optic nerves and spinal cord with relative paucity of cerebral involvement. The typical presentation is with recurrent episodes of extensive spinal cord myelitis with or without recurrent optic neuritis.
MRI findings include paucity of brain lesions, with involvement of 3 or more spinal cord segments/optic nerve enhancement.
CSF findings include polymorphonuclear pleiocytosis without oligoclonal bands.
Anti-aquaporin 4 antibody testing carries a 91% sensitivity and 100% specificity for NMO.
There is extensive evidence to suggest mitoxantrone is superior compared to other therapies in treatment of NMO.

Source:
1. Weinstock-Guttman B, et al. Study of Mitoxantrone for the Treatment of Recurrent Neuromyelitis Optica. *Arch Neurol.* 2006;63(7):957-963
2. Takahashi T, et al. Anti-aquaporin-4 antibody is involved in the pathogenesis of NMO: a study on antibody titre. *Brain* (2007) 130 *(5): 1235-1243*

38. Answer: b

Discussion: The general consensus regarding tobacco alcohol amblyopia is that it is caused by under nutrition with subsequent deficiencies in folate, vitamin B12, thiamine and sulfur-containing amino acids. The papillomacular bundle appears to be particularly susceptible, resulting in progressive dimness of vision, central/centrocaecal scotomas and colour vision defects. Pain is always absent.

39. Answer: a

Discussion: LEMS is a neuromuscular junction disorder associated with malignancy (small-cell lung, stomach, prostate, breast) characterized by autoimmune destruction of voltage gated calcium channels at the pre - synaptic junction with subsequent loss of acetylcholine (Ach) release.
The predominant feature is limb weakness, which is the presenting feature in 95% of patients. This is in sharp contrast to MG where patients commonly present with extraocular muscular weakness. The commonest ocular features are dry eye and abnormal pupil light responses. Extraocular muscular weakness is rare and mild, and when it occurs, also tends to improve with sustained testing in LEMS in contrast to MG.

Source:
1. Wirtz PW, Sotodeh M, Nijnuis M, Van Doorn PA, Van Engelen BG, Hintzen RQ, et al. Difference in distribution of muscle weakness between myasthenia gravis and the Lambert-Eaton myasthenic syndrome. *J Neurol Neurosurg Psychiatry.* Dec 2002;73(6):766-8.

40. Answer: b

Discussion: Sebaceous gland carcinomas are lethal tumors that account for 1% of eyelid malignancies. They occur predominantly in women in their 60s and older. The tumors may mimic chronic blepharitis or recurrent chalazia. The lesions classically involve the upper lid and present as painless, indurated lumps with loss of lashes and lid architecture.

Source:
1. Denniston AKO, Murray PI. 2009. *Oxford Handbook of Clinical Ophthalmology.* 2nd Ed. Oxford. Oxford University Press

41. Answer: d

Discussion: True eyelid colobomas with discontinuity of the eyelid margin are not seen in Treacher Collins Syndrome. Pseudocolobomas arise as a result of an intact lid margin in the presence of a lateral facial cleft that causes infero-lateral dystopia of the lower lid.
True colobomas are seen in Goldenhar syndrome.

Source:
1. Katowitz JA. 2002. *Pediatric Oculoplastic Surgery.* New York. Springer-Verlag.
2. Wilson MA, Saunders RA, Trivedi RH. 2009. *Pediatric Ophthalmology: Current Thought and a Practical Guide.* Berlin Heidelberg. Springer-Verlag.

42. Answer: b

Discussion: These are fairly straightforward questions that do arise from time to time. Particularly, note the extent of horizontal tissue loss as well as which lid is involved.

In general, defects of up to 25% of either lid may be closed directly, or in association with a canthotomy and cantholysis if the defect is slightly larger.

Defects of up to 50% of either lid may be closed with a Tenzel Semicircular flap.

Larger defects of the upper lids may be closed with a Cutler Beard combined anterior/posterior lamellar flap with closure of the eye for several weeks, which may pose difficulties should the contralateral eye have poor vision.

Larger defects of the lower lid may be reconstructed with either a Mustarde's cheek rotation flap and posterior lamellar graft or a Hughes Tarsoconjunctival flap and anterior lamellar graft.

Source:
1. Tyers AG, Collin JRO. 2008. *Colour Atlas of Ophthalmic Plastic Surgery.* 3rd Ed. Elsevier

43. Answer: b

Discussion: Hyphaema is associated with elevated intraocular pressure.

44. Answer: c

Discussion: Senile retinoschisis involves the infero-temporal quadrant in 72% of patients.

Source:
1.Byer NE. Clinical Study of Senile Retinoschisis. *Arch Ophthalmol.* 1968;79(1):36-44.

45. Answer: c

Discussion: As per Royal College guidelines on treatment of diabetic retinopathy, laser photocoagulation should be considered first for patients with clinically significant macular oedema (CSMO) with normal or near normal vision.

Patients with CSMO and a visual acuity of between 24-78 letters should be offered intravitreal ranibizumab as first line treatment, with or without concurrent grid laser.

Patients with CSMO and visual acuity below 24 letters should be examined carefully for evidence of associated macular ischemia which may complicate treatment with ranibizumab.

Source:
1. Royal College of Ophthalmologists Guidelines on Diabetic Retinopathy 2012

46. Answer: c

Discussion: Traditionally, the main contraindication for adjustable sutures would be pediatric strabismus owing to difficulty of patients to co-operate with post surgical adjustment. Adjustable sutures are indicated when there is a need for precise post-operative alignment, specifically in acquired strabismus in binocular patients with troublesome diplopia.

Source:
1. Nihalani BR, Hunter DG. Adjustable Suture Strabismus Surgery. *Eye.* 2011 October 25(10):1262-1276

47. Answer: b

Discussion: Pre and post auricular skin are options for reconstruction of the anterior lamellae of the upper lids. The grafts of choice, however, would be either contralateral upper lid skin or split skin grafts due to them having similar thickness to that of the recipient site.

Source:
1. Tyers AG, Collin JRO. 2008. *Colour Atlas of Ophthalmic Plastic Surgery.* 3rd Ed. Elsevier

48. Answer: b

Discussion: According to Joint BSPAR/RCOphth JIA screening guidelines 2006, risk of flare is greatest in the 6 months post cessation of methotrexate, and children will need to be followed up at 2 monthly intervals for 6 months before reverting to the original follow up schedule.

Source:
1. Joint British Society for Paediatric and Adolescent Rheumatology and Royal College of Ophthalmologists Juvenile Idiopathic Arthritis Screening Guidelines 2006

49. Answer: c

Discussion: Of the mentioned tests, the Wilcoxon Signed Ranks Test is a non – parametric (distribution variable) test designed to compare the means of two dependent data sets.

50. Answer: c

Discussion: Established treatment of adrenal crisis includes steroids in supraphysiologic doses and aggressive fluid therapy to combat hypotension.

Source:
1. Hahner S, Allolio B. Therapeutic management of adrenal insufficiency. *Best Pract Res Clin Endocrinol Metab.* Apr 2009;23(2):167-79

51. Answer: d

Discussion: Risk factors include anterior chamber and sulcus lenses, older age, diabetes mellitus, use of topical latanoprost and uveitis. Examination may reveal blunting of the foveal reflex, parafoveal cystoid spaces and a petalloid hyperfluorescent pattern on fluorescein angiography. The vast majority of cases resolve spontaneously within a year.

Source:
1. Bandello F, Battaglia Parodi M. 2012. *Surgical Retina.* Basel. Karger

52. Answer: b

Discussion: TSPK is characterized by a bilateral, recurrent, focal, intraepithelial keratitis with no conjunctival or stromal inflammation i.e a white, quiet eye. Lesions may consist of up to 50 coarse whitish raised epithelial dots, often within the central cornea, which do not stain with fluorescein.

Source:
1. Tabbara KF, Ostler HB, Dawson C, Oh K. Thyegeson's Superficial Punctate Keratitis. *Ophthalmology.* Jan 1981;88(1)75-7
2. Arffa RC. *Grayson's Diseases of the Cornea.* 4th ed. St Louis: Mosby-Year Book;1997

53. Answer: b

Discussion: Conditions associated with ONH include maternal insulin dependent diabetes as well as midline or hemispheric brain defects of which De Morsier syndrome is the commonest.
Midline bony defects such as basal encephaloceles may also be seen. An MRI is important in all patients with ONH to detect any associated central nervous system abnormality. Endocrinologic consultation is also vital as any existing hormonal deficiency may be life threatening.

Source:
1.Borchert M, McCulloch D, Rother C, et al. Clinical assessment, optic disk measurements, and visual-evoked potential in optic nerve hypoplasia. *Am J Ophthalmol.* 1995; 120:605-612.
2.Brodsky Me. Congenital optic disk anomalies. *Surv Ophthalmol.* 1994;39:89-112.

54. Answer: b

Discussion: Risk factors include immunosuppression, either due to chronic illness, immunosuppressive drugs, or HIV. Increasing age is a risk factor, with patients above 60 years of age at greatest risk. White patients are 4 times as likely to develop HZO as Afro-Caribbean patients.

Source:
1. Schmader K, George LK, Burchett BM, et al. Racial differences in the occurrence of herpes zoster. *J Infect Dis* 1995;171:701-704.
2. Buchbinder SP, Katz MH, Hessol NA, et al. Herpes zoster and human immunodeficiency virus infection. *J Infect Dis* 1992;166:1153-1156.

55. Answer: c

Discussion: Orbital metastases are less common than uveal metastases, the ratio of which has been reported in several case series to be in the region of 1:8. They are rare, comprising less than 10% of biopsy proven orbital neoplasms. In most case series, breast, followed by lung carcinoma, are the most common primary lesions. Diplopia tends to be the most common presenting symptom as reported in multiple case series.

Source:
1.Goldberg RA, Rootman J, Cline RA. Tumors metastatic to the orbit. *Surv Ophthamol* 1990;35:1–24.
2. Burmeister BH, Benjamin CS, Childs WJ. The management of metastases to eye and orbit from carcinoma of the breast. *Aust NZ J Ophthalmol* 1990; 18:187–90.
3. Char DH. *Clinical ocular oncology.* 2nd ed. Philadelphia: Raven-Lippincott, 1996:390–7

56. Answer: d

Discussion: The LGB may be affected by any manner of pathology, including vascular insults, inflammation and tumors. Signs include a normal visual acuity and colour vision (in unilateral lesions – as the ipsilateral hemifields are spared), the absence of an RAPD and the presence of disc pallor.

Retinotopically, the dorsal LGB subserves the macula, lateral LGB the superior field and medial LGB the inferior field. Depending on the extent of the lesion, various homonymous field defects may be seen, which may be congruous or incongruous.
In particular, 2 patterns of field loss are seen with occlusion of the lateral choroidal artery (which gives rise to a homonymous horizontal sectoranopia) and occlusion of the anterior choroidal artery (which gives rise to a homonymous quadruple sectoranopia)
Damage to the adjacent thalamic nuclei or pyramidal tracts may also result in contralateral limb weakness or loss of sensation.

Source:
1. Newman NJ, Miller NR, Biousse V. 2008. *Walsh and Hoyt's Clinical Neuro-Ophthalmology: The Essentials.* 2nd Ed. Lippincott Williams & Wilkins
2. Frisen L. Quadruple sectoranopia and sectorial optic atrophy: a syndrome of the distal anterior choroidal artery. *J Neurol Neurosurg Psychiatry.* 1979 July; 42(7): 590–594.

57. Answer: d

Discussion: Tolosa Hunt Syndrome is caused by a non-specific inflammatory process involving the cavernous sinus and/or the orbital apex and is characterised by severe, 'gnawing or boring' pain behind the globe and ophthalmoplegia and frequently, visual loss.
The episodes spontaneously resolve after an average of 8 weeks.
Recurrences are common, which may occur after a duration of months to years, and while they typically involve the same eye, the contrateral eye may be involved.
The investigation of choice is contrast enhanced MRI of the brain and orbits. Typical findings include enhancing lesions of intermediate density on T1 weighted scans (isointense to muscle). The findings are not pathognomonic, and are also consistent with differential diagnoses of meningioma, lymphoma, or sarcoidosis.

Source:
1. Yousem DM, Atlas SW, Grossman RI, et al. (1990) MR imaging of Tolosa-Hunt syndrome. *Am J Neuroradiol* 10:1181–1184.
2. Kline LB, Hoyt WF. The Tolosa Hunt Syndrome. *J Neurol Neurosurg Psychiatry* 2001;71:577-582

58. Answer: c

Discussion: Stickler's Syndrome is primarily a disorder of Type 2 Collagen.

59. Answer: b

Discussion: Intermittent exotropia is the commonest form of primary exotropia. It presents from ages 2-5, when critical visual pathways have had time to develop in infancy, and thus, most patients have excellent stereoacuity of 50-60 s of arc when their eyes are straight.
Clinical features of progression include photophobia, increasing time of decompensation, and in older children, diplopia.
Surgery is indicated to restore binocular single vision and prevent amblyopia from occurring, usually when the squint is present in 50% or more of waking hours.

Source:
1. Wright KW, Spiegel PH. 2nd Edition. *Pediatric Ophthalmology and Strabismus*. New York. Springer.

60. Answer: a

Discussion: Blue cone monochromatism is an X linked disorder characterised by preservation of blue cone function and partial achromatopsia. The locus Xq28 on which the continuous gene segments for both green and red cones are located is the site for most mutations. Children present at birth with nystagmus, and later in life, with progressive myopia. Visual acuity is often better than 6/60. Fundal examination is usually normal in the 1st decade of life and shows progressive macular atrophic changes with subsequent reduction in central vision in the late teens. Colour vision testing with D15 test reveals protan and deutan defects. ERG reveals a spared rod response with a minimally present cone response.

Source:
1. Brodsky MC. 2010. *Pediatric Neuro-Ophthalmology*. 2nd Edition. Springer

61. Answer: c

Discussion: V patterns are present when the eyes converge more than 15 PD from up gaze to down gaze, an A pattern is present when the eyes diverge more than 10 PD from up gaze to down gaze.
The commonest causes of V patterns include IO over action. Other notable causes of V patterns include craniosynostoses and bilateral superior oblique palsies, which may occur after trauma.
Treatment options for IO over action include IO myotomy.

Source:
1. Lee SY, Cho HK, Kim HK, Lee YC. The effect of inferior oblique myotomy in patients with inferior oblique overaction. *J Pediatr Ophthalmol Strabismus*. Nov 1 2010;47(6):366-72

62. Answer: d

Discussion: Approximately 60% of eyes with Stage 1 macular holes undergo spontaneous vitreo-foveolar detachment with no further progression.

Source:
1. Gass, JDM. Idiopathic senile macular hole: its early stages and pathogenesis. *Arch Ophthalmol* 1988; 106:629-639

63. Answer: b

Discussion: Molecular weight determines the abiity of the viscoelastic to remain in the eye during irrigation. Larger molecules such as sodium hyaluronate, with their longer chains, remain bound to each other, and may exit the eye as a large bolus. Smaller molecules such as chondroitin sulphate demonstrate better retention in the eye, coat surfaces better, and are more difficult to remove. Peak pressure rise from retained viscoelastic occurs within the first 6-8 hours. Viscoelastic aspiration after intraocular lens implantation has been shown to have no beneficial effect on trabecular outflow, but it does reduce the peak and duration of pressure rise.

Source:
1. MacRae SM, Edelhauser HF, Hyndiuk RA et al. The effects of sodium hyaluronate, chondroitin sulphate and methylcellulose on the corneal endothelium and intraocular pressure. *Am J Ophthalmol* 1983;95:332-341

64. Answer: b

Discussion: PPMD is an AD inherited disease that is often innocuous. Physical findings include characteristic posterior corneal vesicles or bands which may occur in association with features similar to ICE syndrome including peripheral anterior synechiae, pupil distortion, corectopia, guttata, and in advanced cases, corneal oedema. However, ICE is almost invariably unilateral as compared to PPMD which is bilateral. (asymmetric)

The condition is rarely symptomatic.

Source:
1. Anderson NJ, Badawi DY, Grossniklaus HE, Stulting RD. Posterior Polymorphous membranous dystrophy with overlapping features of iridocorneal endothelial syndrome. *Arch Ophthalmol.* Apr 2001;119(4):624-5

65. Answer: b

Discussion: Risk factors for pseudoexfoliation syndrome include age, female sex, northern latitudes and higher altitudes

66. Answer: c

Discussion: The abovementioned lesions except metastases are well defined on MRI. Inflammatory or infiltrative lesions tend to be poorly defined on MRI or CT.

67. Answer: b

Discussion: Features suggestive of malignant change within a nevus include change in colour, size, fornix or palpebral location, scleral fixation, corneal extension or increase in number of feeder vessels.
Feeder vessels and cysts are seen in up to 2/3 of patients with naevi that do not otherwise exhibit signs of possible malignancy.

Source:
1. Zembowicz A, Mandal RV, Choopong P. (2010) Melanocytic Lesions of the Conjunctiva. *Archives of Pathology & Laboratory Medicine*: December 2010, Vol. 134, No. 12, pp. 1785-1792.
2. Shields CL, Fasiuddin AF, Mashayekhi A, Shields JA. Conjunctival nevi: clinical features and natural course in 410 consecutive patients. . *Arch Ophthalmol.* 2004 Feb;122(2):167-75

68. Answer: a

Discussion: Pseudoexfoliation glaucoma tends to present unilaterally, with progression to bilaterality in 40% of patients. 20% of patients have associated narrow or occludable angles. Pseudoexfoliation glaucoma tends to be more volatile, and often progressing faster as compared to primary open angle glaucoma. SLT is an established treatment option.

Source:
1. Majka CP, Challa P. *Diagnosis and Management of Pseudoexfoliation Glaucoma.* Ophthalmic Pearls, American Academy of Ophthalmology. www.aao.org/publications/eyenet/200606/pearls.cfm

69. Answer: d

Discussion: Of the above conditions, congenital hereditary stromal dystrophy is likeliest to be visually significant, often requiring penetrating keratoplasty.

70. Answer: c

Discussion: The onset of IK is often unilateral, and eventually becomes bilateral in 80-90% of cases.

71. Answer: a

Discussion: Angioid streaks are associated with breaks in an abnormally calcified Bruch's membrane.

Source:
1. Klien BA. Angioid streaks: A clinical and histopathologic study. *American Journal of Ophthalmology.* 1947;30:955-68

72. Answer: d

Discussion: Of the various sickle cell haemoglobinopathies, HbSS has been estimated to be the most common, affecting 8.5% of the North American population of African descent. HbSThal is much less common, estimated at 0.03% of the population.
Whilst systemic manifestations are most common in patients with HbSS, retinal manifestations occur more frequently in patients with the less common HbSC and HbSThal.
Visual loss in patients with retinal manifestations frequently occurs secondary to proliferative changes with subsequent vitreous haemorrhage.
As autoinfarction occurs in as many as 60% of eyes, treatment is reserved for patients with bilateral disease, extensive proliferation, rapid neovascular growth and/or cases where unilateral loss of vision has already occurred.
Treatment options include direct treatment of feeder arterioles and draining veins (reported to cause regression in 90% of eyes), scatter photocoagulation and cryotherapy.

Source:
1. Farber MD, Jampol LM. Treatment of neovascularization, vitreous haemorrhage, and retinal detachment in sickle cell retinopathy. *Arch Ophthalmol* 1975;80-885-892
2. Jacobson MS, Gagliano DA, Cohen SB et al. A randomized clinical trial of feeder vessel photocoagulation of sickle cell retinopathy. A long term follow up. *Ophthalmology* 1991;98:581-585
3. Lee CB, Woolf MB, Galinos SO et al. Cryotherapy of proliferative sickle cell retinopathy. Part 1. Single freeze thaw cycle. *Ann Ophthalmol* 1975;7:1299-1308

73. Answer: b

Discussion: Keratoconus is a bilateral, non inflammatory corneal ectasia leading to progressive para-axial thinning and steepening with ocular surface distortion and subsequent reduction in vision. Risk factors include allergic conjunctivitis, contact lens wear and frequent eye rubbing. Various associations including retinitis pigmentosa and Leber's Congenital Amaurosis. There is evidence of familial inheritance in up to 15% of patients, and genetic evidence of abnormal superoxide dismutase has been reported.

Source:
□□□Rabinowitz YS. Keratoconus. *Surv Ophthalmol*. Jan-Feb 1998;42(4):297-319
□□□Kennedy RH, Bourne WM, Dyer JA. A 48-year clinical and epidemiologic study of keratoconus. *Am J Ophthalmol*. Mar 15 1986;101(3):267-73.
3. Cristina Kenney M, Brown DJ. The cascade hypothesis of keratoconus. *Cont Lens Anterior Eye*. Sep 2003;26(3):139-46

74. Answer: b

Discussion: Differentials for subretinal solid lesions include naevi, melanomas, melanocytomas, metastases, osteomas, haemangiomas and retinal pigment epithelial (RPE) lesions such as congenital hypertrophy of the RPE. Often, ultrasonography, angiography and to a lesser extent, MRI will aid in reaching a diagnosis.

The ultrasound and angiographic findings strongly suggest a choroidal haemangioma. In contrast, while melanomas may present with macular oedema, ultrasonography typically shows internal hyporeflectivity and early hypofluorescence. Angiographic findings in melanomas may also reveal a dual circulation.
A choroidal metastatic lesion would present with internal hyper-reflectivity on ultrasonography and early hypofluorescence/lack of internal vasculature on angiogram.

Source:
1. Kanski JJ, 2007. *Clinical Ophthalmology: A Systematic Approach*. 6th Ed. Elsevier

75. Answer: c

Discussion: Bell's palsy is the most common cause of unilateral lower motor neurone 7th nerve palsy. Features include involvement of the frontalis and lower facial muscles. Brow ptosis, lagophthalmos and drooping of the lower lid with subsequent pump failure and epiphora are often seen. Hyperacusis and impaired taste is often present.

Source:
1. May M. 2000. *The Facial Nerve*. Thieme Medical.

76. Answer: d

Discussion: The scenario suggests a post-ganglionic Horner syndrome. Of the listed options, only option 'd' may be a lesion involving the 3rd order neuron in the sympathetic chain.

77. Answer: d

Discussion: The average length of the neonatal eye has been reported to be 17-17.2 mm long

Source:
1. Warburg M. Classification of microphthalmos and coloboma. *J Med Genet* 1993;30:664-669

78. Answer: d

Discussion: Choroidal naevi do not demonstrate autofluorescence. Choroidal melanomas, which frequently contain lipofuscin deposits, may demonstrate autofluorescence.

79. Answer: d

Discussion: HLA B27 related uveitis tends to occur in young males. The commonest manifestation is a unilateral AAU with severe inflammation. Fibrin is seen in 25%, hypopyon in 17% and there is an increased risk of posterior synechiae as compared to idiopathic AAU. Recurrences are more common than in idiopathic disease. HLA B27 related uveitis more commonly affects men.

Source:
1. www.uveitis.org

80. Answer: b

Discussion: SFEMG is more sensitive than RNS, with reported sensitivities of 100% and 44-65% respectively. In addition, SFEMG abnormalities are not masked by pyridostigmine.
In SFEMG for ocular MG, the tested muscle of choice is the frontalis, and in systemic MG, the extensor digiti communis. The commonest stimulation rate used in RNS is 3 Hz, and any abnormalities may be masked by concomitant use of pyridostigmine

Source:
1. Padua L, Stalberg E, LoMonaco M, Evoli A, Batocchi A, Tonali P. SFEMG in ocular myasthenia gravis diagnosis. *Clin Neurophysiol.* Jul 2000;111(7):1203-7

81. Answer: b

Discussion: This is likely to be a case of idiopathic intracranial hypertension, given the symptoms, bilateral disc swelling, normal CT scan and raised LP opening pressure. MRI and MRV are progressively becoming the investigation of choice, as they are more sensitive for meningitis and dural venous sinus thrombosis that should be excluded, given a history of contraceptive use and a normal BMI.

Source:
1. Gans, MS. Idiopathic Intracranial Hypertension. *emedicine.medscape.com*

82. Answer: b

Discussion: The abovementioned scenario would suggest a provisional diagnosis of orbital cellulitis. However, early involvement of a previously normal contralateral eye would suggest a diagnosis of cavernous sinus thrombosis.
MRV is the imaging method of choice to detect lack of flow within the sinus.

Source:
1. Chiewvit P, Piyapittayanan S, Poungvarin N. Cerebral Venous Thrombosis: Diagnostic Dilemma. *Neurol Int.* 2011 November 29;3(3):e13

83. Answer: a

Discussion: Conjunctival melanocytic tumors should be referred if they demonstrate nodularity, feeder vessels, or are located at the caruncle, fornix or inner surface of the lids. Clear cysts are not a sign of malignancy and are not a criterion for referral.

Source:
1. The Royal College of Ophthalmologists *Referral* Guidelines for adult ocular *tumours* including choroidal naevi. October 2009

84. Answer: c

Discussion: Corneal laceration wounds should be closed ideally with a 10 or 11-0 nylon suture swaged to a spatulated 3/8 needle. Wounds should be closed first, at easily identifiable landmarks, for example, the corneal limbus. Needles should enter at a 90 degree angle to the corneal surface to achieve a 90% depth bite Sutures should be longest at the limbus and shortest closest to the centre of the cornea.

Source
1. Ferenc Kuhn. 2008. *Ocular Traumatology.* Springer.

85. Answer: b

Discussion: As per European Group On Graves Orbitopathy (EUGOGO) guidelines, sight threatening thyroid eye disease should be treated promptly with pulsed intravenous methylprednisolone – which has shown to have higher efficacy and less adverse events if compared to oral or retrobulbar steroids.

Source:
1. Bartalena L, Baldeschi L, Dickinson AJ, Eckstein A, Kendall-Taylor P, Marcocci C, et al. Consensus Statement of the European Groupon Graves' Orbitopathy (EUGOGO) on Management of Graves' Orbitopathy· *Thyroid*. Volume 18, Number 3, 2008

86. Answer: c

Discussion: Downbeat nystagmus is a conjugate, involuntary ocular movement characterised by a slow, defoveating, upward pursuit movement followed by a fast, corrective downward saccade.
Lesions usually localise to the cerebellum (up to 88%) but may be due to lesions within the brainstem. The classic cause in children is Arnold-Chiari malformation that causes pressure on the craniocervical junction, but may be caused by infarction, drug toxicity, degenerative syndromes and demyelination involving the cerebellum and pons.

Source:
1. Yee RD. Downbeat nystagmus: characteristics and localization of lesions. *Trans Am Ophthalmol Soc*. 1989; 87: 984–1032.

87. Answer: d

Discussion: Neovascular AMD typically may reveal hyperfluorescence due to leak from the choroidal neovascular complex.

88. Answer: d

Discussion: The above scenario is likely to point to a case of acute demyelinating optic neuritis. Diagnosis is clinical, but an MRI with a FLAIR sequence is of help in prognosticating the risk of eventually developing multiple sclerosis.

89. Answer: a

Discussion: HIV is a single stranded RNA virus of the Retroviridae family, genus Lentivirus. Both HIV-1 and 2 may cause human infection. Co-infection with HPV, Hepatitis B and C are common. Global distribution differs greatly, with the more aggressive HIV-1 virus seen in the developed world. HIV-2 is seen predominantly in the developing world, carries a lower risk of transmission, and tends to progress more slowly to AIDS as compared to HIV-1. Significance of this global disparity is that most research and development tends to be focused on vaccines for HIV-1.

Source:
1. Robertson DL, Anderson JP, Bradac JA *et al.*: HIV-1 nomenclature proposal. *Science* 288, 55-56 (2000).
2. Eholie S, Anglaret X: Commentary: decline of HIV-2 prevalence in West Africa: good news or bad news? *Int. J. Epidemiol*. 35, 1329-1330 (2006).

90. Answer: a

Discussion: Hashimoto's disease presents with thyroid destruction and fibrosis, causing low free T4 levels and increasing TSH levels. TP antibodies are seen in 90% of patient's with Hashimoto's whereas TSH receptor antibodies are rarely positive.

Printed in Great Britain
by Amazon